TABLETOP MATH GAMES COLLECTION

More 49 Ways to Play Math

MATH GAMES FOR ALL AGES

VOLUME TWO

Denise Gaskins

Tabletop Academy Press

Contents

Learning Math Through Play...5

Mental Math Do's and Don'ts...6

Journaling With Games...7

Childhood Classic Games...9

Bango...11

Crazy Eights..13

Wild and Crazy Eights..15

Michigan (Boodle)...17

Fan Tan (Sevens)...21

Fan Tan Variations..23

Dominoes..25

For More Information...28

Number Bond Games...29

Odd One Out..31

Domino Bond...33

Nine Cards..35

Tens Go Fish...37

Shut the Box...39

Shut the Box Variations..43

I Have, You Need...45

For More Information...46

Copyright © 2024 Denise Gaskins
All rights reserved.
Tabletop Academy Press, Boody, IL, USA, TabletopAcademyPress.com.

EDUCATIONAL USE: You have permission to copy the math games in this book for personal, family, or single-classroom use only. Please enjoy them with your homeschool students, math circle, co-op, or other small local group.

ABOUT THE FONT: The Cadman font was designed by P.J. Miller to be as reader-friendly as possible. Letters are distinct and easy to distinguish, especially those most often confused by children and adults with dyslexia.

Games with Numbers to 100 47
   Thirty-One 49
   Push the Penny 51
   Dollar Nim 57
   Countdown 59
   Euclid's Game 61
   Number Grid Tic-Tac-Toe 63
   For More Information 68

Yan Tan Tethera 69
   The Joyful Challenge of Mental Exercise 71
   Yan Tan Tethera 72
   How To Play 73
   Sample Game 74
   Yan Tan Tethera Gameboards 75
   For More Information 80

Games for Modeling Multiplication 81
   What Is a Math Model? 83
   Multiplication Model Cards 84
   Twelve Cards 91
   Go Fish 93
   Concentration (Memory) 95
   Multiplication Rummy 97
   Multiplication Number Train 99
   For More Information 100

Times Table Games 101
   Galactic Conquest 103
   Galactic Conquest Variations 107
   Domino Product Train 109
   Times-Tac-Toe 111
   Multiplication Gomoku 115
   The Product Game 119

- Ultimate Multiple Tic-Tac-Toe 125
- For More Information 128

The Great Escape 129
- The Great Escape 131
- Secret Number Code 132
- Team Great Escape 135
- For More Information 140

Integer Games 141
- Consecutive Capture 143
- Strike It Out 145
- Integer Solitaire 147
- Grid Fight 149
- Honeycomb 153
- Four in a Row Integers 159
- For More Information 164

Mental Math Mastery 165
- Operations 167
- Exponent Pickle 171
- Krypto 175
- Power Krypto 179
- Krypto Insanity 181
- Equation Master 185
- Four in a Row Algebra 187
- For More Information 192

Algebra Match 193
- Algebra Match 195
- Algebra Match Variations 197
- Sample Algebra Match Cards 198

Special Thanks 202

Playful Math Books by Denise Gaskins 203

Learning Math Through Play

Clear off a table, find some dice or a deck of cards, and you're ready to enjoy playing math. Keep these tips in mind:

- Game rules are a social convention, easy to change by agreement among the players. Feel free to invent your own rules, and encourage players to modify the games as they play. As they tinker with the game, it prompts them to think more deeply about the math concepts involved.

- Rating math games by grade level is inherently arbitrary. Young players may eagerly join a game with advanced concepts if the fun of the challenge outweighs the work involved. On the other hand, don't worry that a game is too easy for anyone—even adults—as long as they find it interesting. Everyone benefits from a little extra practice in math, but it's the logic of strategy that makes a game fun.

- If you are a parent, these games provide opportunities to enjoy quality time with your children. If you are a classroom teacher, use the games as warm-ups and learning center activities or for a relaxing review day at the end of a term. If you are a tutor or homeschooler, make games a regular feature in your lesson plans to build your students' mental math skills.

- Try to let people learn by playing. Explain the rules as simply as possible and jump into the fun. You can add details, exceptions, and special situations as they come up during play or before starting future games.

- Be warned: Although children can play most of these games on their own, they learn much more if adults play along. When adults join the game, they reinforce the value of mathematical play.

- Talk as you play, especially with kids. As you watch your children's responses and listen to their comments, you'll discover what they understand about math. Where do they get confused? What do they do when they're stuck? Can they use the number patterns they've mastered to figure out something they don't know? How easily do they give up?

  Real education, learning that sticks for a lifetime, comes through person-to-person interactions. Children absorb more from the give and take of discussion with an adult than from even the best workbook or teaching video.

Mental Math Do's and Don'ts

Math games stretch young players' abilities to manipulate numbers in their heads. But please don't treat these games as worksheets in disguise. A game should be voluntary and fun. No matter how good it sounds to you, if a game doesn't interest your family, put it away. You can always try another one tomorrow.

You'll know when you find the perfect game because your children will wear you out wanting to play it again and again and again.

As you play, remember these tips for mental calculation:

- Don't just count. Limit straight counting to a few steps, so you can't lose track. That means you may work 39 + 2 by counting, but not 39 + 7. When you do count, always start at the bigger number, so you have fewer steps.

- Do break numbers apart. Work with the easier parts first.

- Do use logic to rearrange your numbers and simplify your calculation. For instance, to find 39 + 7 you can imagine moving one piece from the seven to the big pile: 39 + 7 is the same as 40 + 6.

- Don't try to keep track of "borrowing" or "carrying" numbers while you work. But you can use funny numbers as an intermediate step. For example, if you remember that 9 + 7 = 16, then you might think of 39 + 7 as thirty-sixteen.

- It often helps to add or subtract more than you really need, then adjust the calculation to fit. So instead of trying to add 39 + 7, you can add 40 + 7 and then take away the extra one.

- Don't try to memorize everything. Do memorize a few basic facts (such as the pairs of numbers that make ten) that you can use to figure out other things.

- Don't use the same trick on every calculation. Be creative, looking for new ways to use number relationships as you figure things out.

- Do use fingers, manipulatives, or marks on paper to keep track of information while you work, especially with longer, multistep calculations.

- Do allow your children plenty of time to think. Don't worry if a child stares blankly into space. That's often what "thinking hard" looks like. Try not to break their concentration.

Journaling With Games

Games are the ultimate re-playable activity prompts. As you repeat a game, you can try variations on your previous moves to gain extra advantage. This mirrors the approach a mathematician may take when faced with a problem. What if we try this, or that? How do things change, and what stays the same?

After you master the ordinary version of a game, try a *misère* variation. In a *misère* game, the move that otherwise would win now makes you the loser. Players must reconsider their strategy and think more deeply about the game. Consider other ways to modify the game rules. Write about your ideas:

♦ If the game uses playing cards, can you figure out a way to play it with dice or dominoes? Or transfer it to a gameboard?

♦ What if we changed the number of cards to draw, or how many dice to throw on each turn?

♦ Or is there a way to use money in the game? Or can you change it into a whole-body action game? Perhaps using sidewalk chalk?

Older players may want to analyze a game, which can make a great writing prompt. What do you notice about the game, and what does it make you wonder?

♦ Does one player have the advantage, or do both players have an equal chance of winning?

♦ What's the best move? Can you find a strategy to increase your odds?

♦ Are fairness and randomness linked? Why or why not?

Unschooling advocate Pam Sorooshian explains the connection between games and math this way:

> "Mathematicians don't sit around doing the kind of math that you learned in school. What they do is 'play around' with number games, spatial puzzles, strategy, and logic. They don't just play the same old games, though. They change the rules a little, and then they look at how the game changes.
>
> "So, when you play games, you are doing exactly what mathematicians really do—if you fool with the games a bit, experiment, see how the play changes if you change a rule here and there. Oh, and when you make up games and they flop, be sure to examine why they flop—that is a big huge part of what mathematicians do, too."

All of the games in the *Tabletop Math Games Collection* include a matching page for writing your notes and modifications.

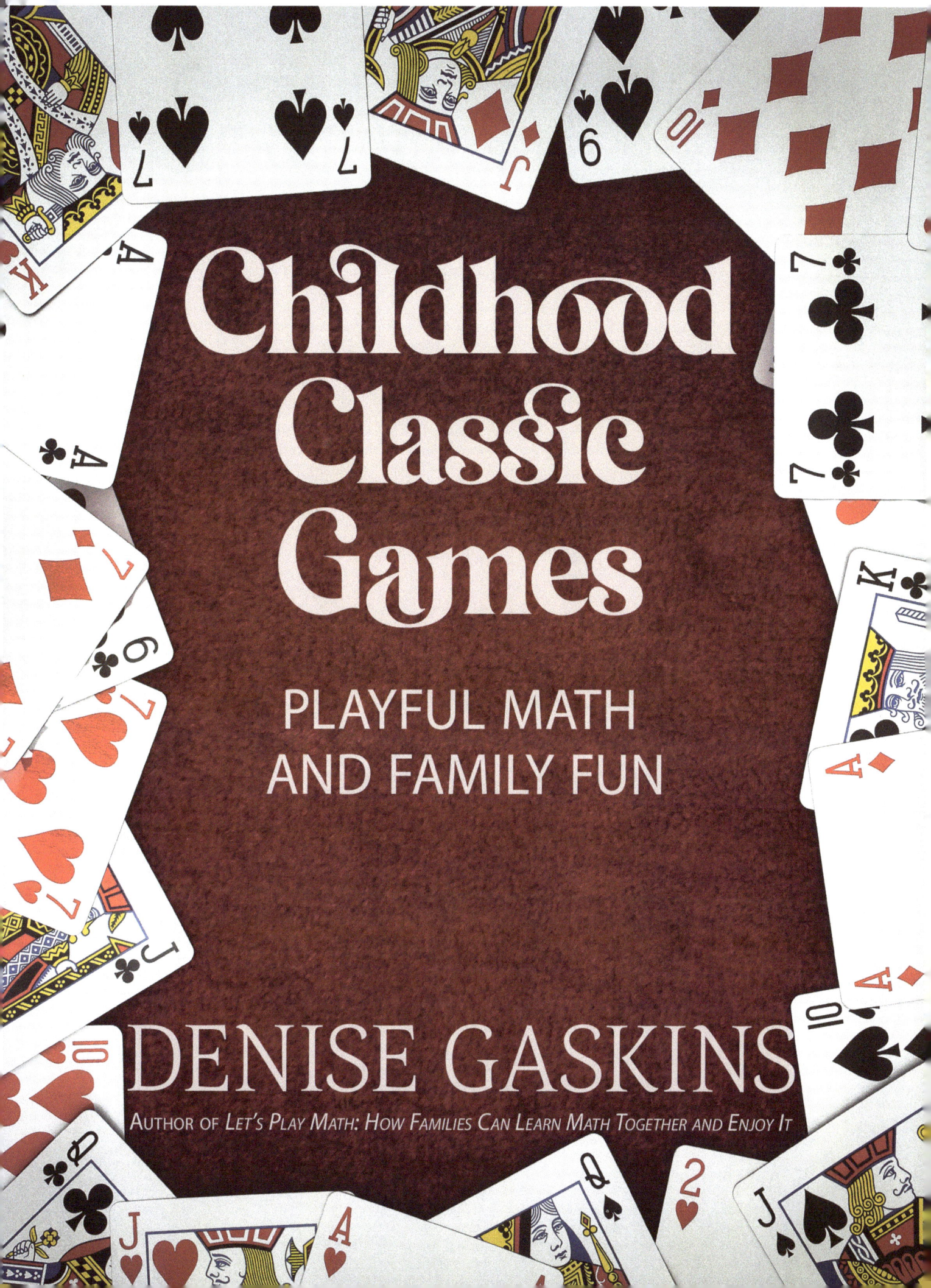

Notes & House Rules

Bango

MATH CONCEPTS: number symbols.
PLAYERS: three or more.
EQUIPMENT: two decks of playing cards.

How to Play

With the first deck, deal five (or more) cards to each player except the dealer. Players arrange their cards face up in one or more rows.

Then the dealer shuffles the other deck of cards and turns up one at a time, reading its value (suit doesn't matter). Any player who has a matching card turns it face down, but a player may turn only one card down each time.

The first player to turn down all five cards calls "Bango!" and wins the right to deal the next round.

Variation

BANGO-TAC-TOE: Deal nine cards to each player, who must arrange them in three rows of three cards each. Or deal sixteen cards, which make four rows of four. The first player to turn down all the cards in a row (vertical, horizontal, or diagonal) wins that round.

You may arrange your Bango cards face up however you like.

Notes & House Rules

Crazy Eights

MATH CONCEPTS: identifying and matching card attributes, wild cards, thinking ahead.
PLAYERS: two or more. More is better.
EQUIPMENT: one deck of playing cards, or a double deck for four or more players.

How to Play

Deal eight cards to each player. Place the rest of the deck face down in the middle of the table, and turn the first card face up beside it to start the discard pile. If this card is an eight, shuffle it back into the deck and turn up the new top card.

On your turn, play a card that matches an attribute (suit or number or face card rank) of the most recent discard. Or you can play an eight, which is a wild card and doesn't have to match anything. Whenever you play an eight, you get to name a new suit for the next play.

If you don't have a matching card or an eight to play, you must draw until you find one. When the draw pile is depleted, the dealer shuffles the discard pile (except for the top card) to make a new draw pile.

The first player to run out of cards wins the hand. If you are keeping score, the other players add up the cards they are holding, as follows.

- ♦ 50 points for each eight
- ♦ 10 points for each face card
- ♦ Face value for each number card

Whoever has the lowest score at the end of the party wins the game.

Variations

The most common variation is that players must announce when they are playing their next-to-last card. If you forget, you must draw two penalty cards.

For a shorter game, you're allowed to pass if you can't play after drawing two cards.

Jokers can serve as additional wild cards, acting like eights.

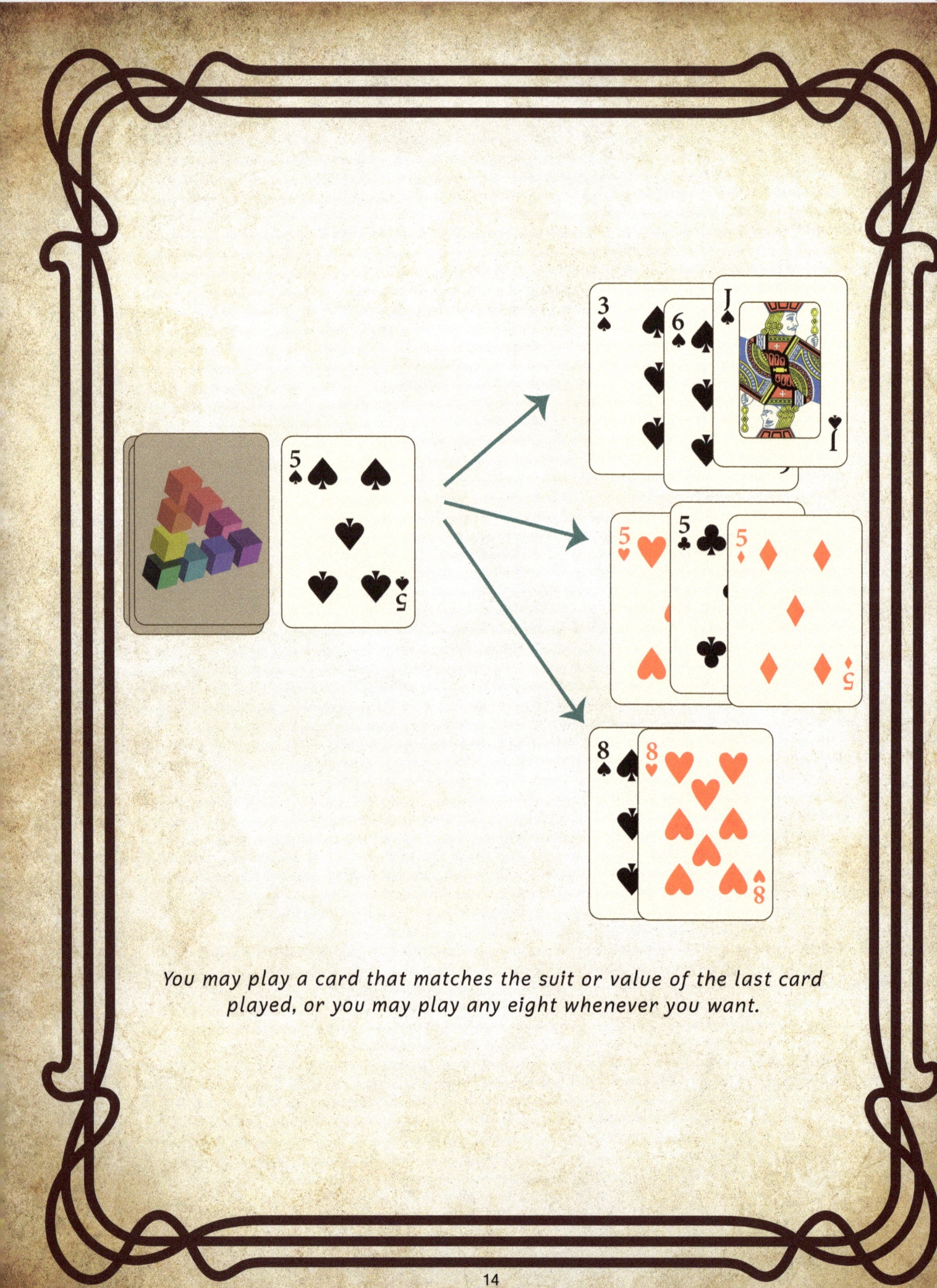

You may play a card that matches the suit or value of the last card played, or you may play any eight whenever you want.

Wild and Crazy Eights

Instead of keeping score, let the winner of each hand add a new rule, which applies to all future hands until the party breaks up. Rules must apply equally to all players. For example:

- Jacks (or other designated cards) reverse the direction of play.
- When a two or four is played, the next player must draw that many cards and does not get to discard.
- Sevens skip the next player.
- When someone plays a queen, everyone passes their hand to the next player. The person who played the queen may choose whether to pass the cards left or right.
- Play know-it-all style: all players lay their cards face up on the table.
- All cards must be played left-handed. Players who forget must draw two cards (using their left hand, of course).
- Playing an eight cancels a draw or skip card.
- If you play an ace, you get to give one card from your hand to any other player.
- Prime time: if you play a prime number, you get a free turn.
- Allow runs: if you can match the number or rank of the card on the discard pile, you may continue to lay down as many additional cards as you wish, according to normal matching rules. (No eights!)
- Claim jumping: if you have a card that exactly matches the card just played, you may slap it down. Play continues on from your position, skipping the intervening players.
- If you get down to one card, you must recite a (short) poem.
- If you play a king, you may swap hands with the player of your choice.
- If you play an eight, you must tell a secret—not a serious secret, but something the other players would not be likely to know.

Instead of making a new rule, the winner of a hand may unmake an earlier one. This option should be used with care, however, since it's the piling on of silly rules that makes the game fun.

Notes & House Rules

Michigan (Boodle)

MATH CONCEPTS: numerical order, sorting by attribute (card suits), standard rank of playing cards (aces high), thinking ahead.
PLAYERS: three or more, up to as many as fit around your table.
EQUIPMENT: one complete deck of cards (including face cards), plus four boodle cards from a second deck; small prizes to go on the boodle cards. Provide a card holder for young children.

Set-Up

Place the boodle cards (also called pay cards) face up in the middle of the table: ace of hearts, king of clubs, queen of diamonds, jack of spades. Or use the Boodle-card page as a gameboard.

Put a prize on each pay card, and add another prize each time a new round is dealt. The boodle prizes might be a coin, a piece of candy, a collectible sticker, or anything else that seems appropriate.

Or you can keep score with poker chips. Then the dealer places two chips on each Boodle card (for a total of eight chips), while the other players put one chip on each of the cards—and you must play enough rounds that everyone takes a turn as dealer.

You can make a card holder by stapling together pieces of stiff cardboard, or two plastic lids stacked top-to-top.

Decorate with stickers, if desired.

Young children slip playing cards inside to hold them spread-out, easy to see.

How to Play Boodle

Remind everyone that in this game, aces are the highest cards in each suit. The dealer deals out all the cards one at a time, one hand to each player and an extra hand called the spare. (It doesn't matter if some players end up with one more card than the others have.) Players may look at their own cards, but the spare hand is left face down, out of play.

The player to the dealer's left begins by laying down (face up) her lowest card in any suit—it doesn't have to be the lowest in her hand, just in that suit—and saying its name out loud. Whoever has the next higher card in that suit can play, and then the next, with the players putting their own cards face up on the table in front of them as they say the names. The cards are not all mixed together in a discard pile.

Continue until no one can play the next higher card (it may be in the spare hand or have been played earlier) or until someone plays the ace to top out that suit. Then whoever played the last card can start a new run. Like the first player, he may play any suit, but it must be the lowest card he has in that suit.

In the course of play, if you lay down one of the pay cards, you get to claim that boodle prize. Any prize not claimed stays on the boodle, as a bonus for the next hand.

As soon as any player runs out of cards, the play ends. If you are playing for poker chips, then all the other players count their remaining cards and pay that many chips to the player who went out.

Variations

Players must change suit when starting a new run. If the person who played the last card cannot change suit, play would pass to the left until someone can. If none of the players has a different suit, then the hand is done and the remaining boodle prizes are left unclaimed.

PLAY THE SPARE: Before play begins, the dealer may choose to discard his hand and pick up the spare, but he must decide without peeking at the spare cards. If you are playing for chips, then the dealer may instead sell the spare hand to the highest bidder, who must pay the dealer (in chips) and discard her original hand before picking up the spare.

Notes & House Rules

Fan Tan (Sevens)

Math Concepts: sorting by attribute (card suits), counting up, counting down, standard rank of playing cards (aces low).
Players: two or more, best with four to six.
Equipment: one complete deck of cards (including face cards), or a double deck for more than six players. Provide a card holder for young children.

How to Play

Deal out all the cards, even if some players get more than others. The player to the dealer's left begins by playing a seven of any suit. If that player does not have a seven, then the play passes left to the first player who does.

After that, on your turn you may lay down another seven or play on the cards that are already down. If you cannot play, say, "Pass."

Once a seven is played in any suit, the six and the eight of that suit may be played on either side of it, forming the fan. Then the five through ace can go on the six in counting-down order, and the nine through king can go on the eight, counting up. You can arrange these cards to overlap each other so the cards below are visible, or you can square up the stacks so only the top card is seen.

Players do not need to wait for both the six and eight of a suit to be played before building the fan up or down.

The first player to run out of cards wins the game.

If you want to keep score, count the cards remaining in your hand after one player goes out. After everyone has had a turn as dealer, whoever has the lowest total score is the champion.

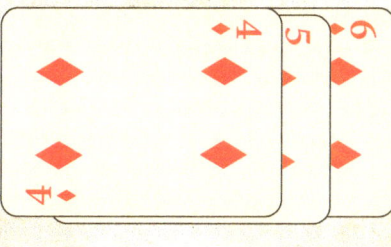  

A Fan Tan game in progress.

Notes & House Rules

Fan Tan Variations

In some traditions, play always begins with the seven of diamonds, so whoever has that card goes first.

Domino Tan

The player to the dealer's left may lead any card, and then all the suits must start with that number (instead of with seven) and build up and down from there.

Fan Tan Trumps

When the dealer gets to the end of the deck and there aren't enough cards to give every player one more, the last few cards are turned face up and may be played by anyone as needed. The suit of the last card becomes the trump suit, and cards of that suit may be played on any of the fans, with the card they replace going on the trumps fan. In this case, the cards must be laid out in overlapping rows, not stacked up, so everyone can see where the trumps have gone.

For instance, if spades are trump, then a nine of spades could be played on the eight of hearts, which would leave the nine of hearts without a home—so it has to go on the spades fan.

EXCEPTIONS: The seven of the trump suit starts its own fan, like any other seven, and the last card dealt (the one that named the trump suit) must also be played to the trumps fan when its turn comes.

Crazy Tan

Deal only seven cards to each player, and set the rest of the deck out as a draw pile. The first player who cannot play must draw one, which he may play if possible. If not, and the next player also cannot play, she must draw two. If neither of those cards will play, and the next player has nothing to play, he must draw three, and so on, each player drawing one more card than the last person.

When one of the players is finally able to lay down a card, this resets the draw count back to zero.

In Crazy Tan, players are allowed to lay down a run (playing several cards in a row of the same suit on a single turn). Or they may play parallel cards (cards of the same rank in different suits, all played in the same turn). Or a player may even lay down parallel runs, if the cards happen to work out that way.

Notes & House Rules

Dominoes

MATH CONCEPTS: subitizing dot patterns, thinking ahead.
PLAYERS: two or more.
EQUIPMENT: one set of double-six or double-nine dominoes, or two sets for five or more players.

Set-Up

Turn all domino tiles (also called bones) face down on the table and mix them around. Each player draws several tiles, as follows:

- Two players draw seven tiles each.
- Three or four players draw five tiles each.
- Five or more players use a double set of dominoes and draw five each.

Set your tiles upright on their sides, so the other players cannot see them. These tiles are your hand, and the ones left on the table are called the wood pile (also known as the bone yard).

How to Play

Whoever has the highest double goes first, placing that tile face up on the table. If no doubles are available, turn everything face down, reshuffle, and draw again.

Play proceeds to the left around the table. Domino tiles are played end-to-end in a long row (the train), with only the outer ends available for adding new tiles. On your turn, you may play one of the tiles in your hand to either open end of the train if the dots (called pips) on one side of your tile match the tile on that end.

Doubles are placed crosswise to the direction of the train, with the middle of the domino touching its neighboring tiles. Players may not match tiles to the ends of the doubles, however, only to the middle of the other side, continuing the train in whichever direction it was growing.

If you have no tile that will play, draw one tile from the wood pile. If you can play that one, do so. Otherwise, add it to your hand. This marks the end of your turn. If you cannot play from your hand and there are no tiles left to draw, you must pass.

The first player to run out of tiles wins the game (or in Muggins, wins that round). If no player is able to go out, then all players add up the pips on their remaining tiles. The player with the smallest sum wins.

The domino train can turn corners and snake around the table as you play.

Domino Variations

Domino games vary tremendously around the world, and even from one family to another within the same town. Whenever you play with friends, be sure to agree on the rules before you draw the first tile.

Cross Dominoes (Sebastopol)

Make a double train. After the first double tile is played, the next four tiles must match it. Two of these are placed normally, at the middle of each side of the first tile, and the other two connect to the ends.

No other tiles may be played until this center cross is formed. Then play continues normally, except in four directions instead of two.

Muggins

Play as for Cross Dominoes, but keep score as you go along. At the end of each turn, the player adds up the pips showing on the live tiles at all four ends of the train, including both halves of any exposed doubles. If these make a multiple of five, the player adds that number to his or her score.

At the end of each round, players add up the pips remaining in their hands, round to the nearest five, and subtract those points from their total so far. The first player to reach 200 points (or some other agreed-upon total) wins the game.

Fives and Threes

Play as for Muggins, but score all multiples of three or five.

Falling Tiles

Forget the gameplay and have fun standing domino tiles in a row and then knocking them down.

For More Information

Bango

Also known as Card Bingo, Bango is a simplified version of the traditional five-in-a-row game Bingo. John McLeod offers more Bingo variations at Pagat.com, a wonderful site for learning about card games from around the world.

- ♦ pagat.com/banking/bingo.html

Fan Tan

Fan Tan may also be called Crazy Sevens. Like any folk game, it is played by a variety of rules around the world. If you search for it on the Internet, you may run into an unrelated Chinese gambling game called Fan Tan, which is similar to Roulette.

Dominoes

Domino-like tile games seem to have originated in China, and they came to Europe through the great trading cities of Venice and Naples.

Some game historians claim the European game was invented independently, because European domino sets are different from Chinese sets in several ways. For instance, Chinese tiles come in suits, like a set of playing cards.

Dominoes spread across France and reached England in the late eighteenth century, where the game became a favorite pastime in British pubs.

A Domino Puzzle

Encourage players to examine a set of domino tiles and describe what they notice. For example, every possible combination (double-0, 0|1, 0|2, etc.) is a single tile, but there are no duplicates: 0|1 is the same tile as 1|0.

Ask them, "If you bought a set of dominoes at a garage sale, how could you tell whether any of the tiles were missing? Can you figure out how many tiles there should be?"

♦ ♦ ♦

Spoiler: To find the answer, make a systematic list, and be careful not to count any of the combinations twice. A double-six set should have twenty-eight tiles, and a double-nine set will have fifty-five. A new set from the store may contain extra blank tiles, which can be decorated with paint or white nail polish to replace lost pieces.

TABLETOP MATH GAMES COLLECTION

Number Bond Games

7 Ways to Play Math with Young Learners

Notes & House Rules

Odd One Out

Math Concepts: simple addition, number bonds for five.
Players: two or more.
Equipment: one deck of playing cards, or a double deck for three or more players.

How to Play

Remove all numbers greater than five from the deck, and add a single face card. For young hands, be sure to provide a card holder.

Shuffle well and deal out all the cards. It does not matter if some players get one more card than the others. All players remove from their hands any fives and any pairs of cards that add up to five. Lay these cards face up on the table, so everyone can see that the discarded pairs match.

On your turn, you may ask any other player for a card. That player must fan out his or her cards as much as possible and hold them so you cannot see the numbers. Then you may choose any card you wish. If the card you take will combine with one of the cards in your hand to make a sum of five, discard the pair. Otherwise, place the card in your hand.

When all the cards are paired and set aside, the player left holding the face card loses the game.

Variations

At our house, we take from the player to our right, so each player first gets a card and then gives one to the next player.

You can play with other number bonds, collecting pairs that add up to eight, or to ten. Whichever number you choose, remove all higher numbers from the deck.

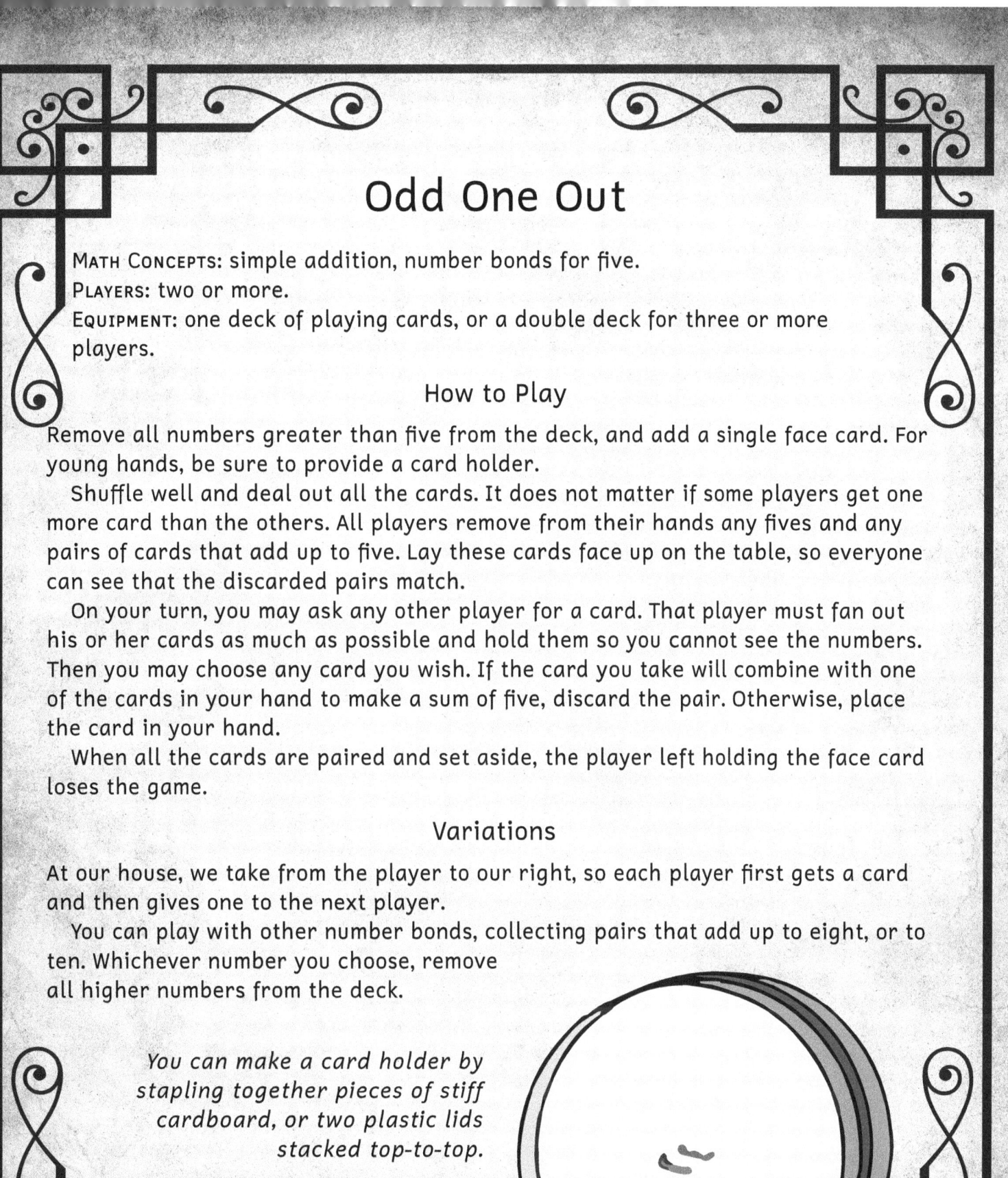

You can make a card holder by stapling together pieces of stiff cardboard, or two plastic lids stacked top-to-top.

Decorate with stickers, if desired.

Young children slip playing cards inside to hold them spread-out, easy to see.

Notes & House Rules

Domino Bond

Math Concepts: addition, number bonds for six.
Players: two or more.
Equipment: one set of double-six dominoes.

Set-Up

Turn all domino tiles face down on the table and mix them around to make the wood pile. Each player draws several tiles, as follows:

- Two players draw seven tiles each.
- Three or four players draw five tiles each.
- Five or more players use a double set of dominoes and draw five each.
- Set your tiles upright on their sides, so the other players cannot see them.

How to Play

Whoever has the highest double goes first, placing that tile face up on the table. If no doubles are available, turn everything face down, reshuffle, and draw again.

On your turn, you may play one of the domino tiles in your hand to either open end of the train, if the pips on one side of your tile and on the end of the train add up to six. For instance, a two can be played next to a four, or a blank next to a six.

Doubles are placed crosswise to the domino train, but the next tile connects to the other side (not the ends) of the double and continues building the train in the same direction. If the train needs to turn because of table space, you may play off any side of the previous tile, pointing your tile in the new direction.

If you have nothing that will play, you may draw one tile from the wood pile. If you can play that one, do so. Otherwise, add it to your hand. This marks the end of your turn. If you cannot play from your hand and there are no tiles left to draw, you must pass.

The first player to run out of tiles wins the game.

Variations

Fives Dominoes: For very young players, remove all tiles with sixes on them. Play by matching tiles that add up to five.

Tens Dominoes: Use a double-nine set of dominoes, removing all the tiles with blanks. Begin play with the double nine or the largest available double tile. Match numbers that add together to make ten.

Notes & House Rules

Nine Cards

Math Concepts: addition, number bonds for ten.
Players: two or more.
Equipment: one deck of playing cards, face cards and jokers removed.

How to Play

The first player shuffles the deck and then turns up the top nine cards, placing them face up in a 3 × 3 array: three rows with three cards in each row. The player captures (removes and keeps) any tens and any pairs of cards that sum to ten, then passes the deck to the next player.

Each player in turn deals out enough cards to fill in the empty spots in the array—or, if there are no empty spots to fill, the player covers all nine cards with new ones. Then capture any tens and any pairs that make ten, and pass the deck on.

Bonus: If your capture uncovers an older card, you may also look for a pair to match that one.

The game ends when the deck is gone or when there are not enough cards left to fill in the holes in the array. The player who dealt last may finish that turn.

Whoever has collected the most cards wins the game.

Variation

Concentration (Memory): Lay all the cards out face down on the table in a single layer with no overlaps. On your turn, flip up two cards. If you find a ten or a number bond, take it. If not, leave the cards showing long enough that all the players can see what they are. Then turn them face down before the next player's turn.

Claim the ten and the pairs of cards that add up to ten.

Do not take longer sums, like 5 + 2 + 3.

Notes & House Rules

Tens Go Fish

Math Concepts: addition, number bonds for ten.
Players: two or more.
Equipment: one deck of playing cards, face cards and jokers removed.

How to Play

Deal seven cards to each player, or five cards if you have four or more players. Put the remaining cards face down in the center of the table and spread them out to make a roughly circular fishing pond.

Players look at their own hands and set aside any tens and any pairs that add up to ten. Each player creates a personal score pile, or fish basket.

On your turn, you may ask one other player, "Do you have a ____?" The blank is for a number that will pair with one of the cards in your hand to make a sum of ten.

For instance, if you have a three, you might ask, "Do you have a seven?" If there are more than two players, the request must be addressed to a specific person. You may ask, "George, do you have a seven?" but it is illegal to say, "Does anyone have a seven?"

- If George has the card you want, he must give it to you.
- If not, he says, "Go Fish." Then you draw any card you wish from the fishing pond.
- If you draw a ten, add it to your fish basket and draw again. If you draw a card that makes a pair of ten with a card in your hand, add that pair to your fish basket. Otherwise, add the card to your hand.

If the fishing pond goes dry, every player who has two or more cards in hand places one card face down to replenish the stock. Mix these cards thoroughly before continuing the game.

The game is over when one player runs out of cards. The other players throw their remaining cards into the fishing pond. (Those fish were too small to keep.) Then all players count the cards in their fish basket pile. Whoever caught the most fish wins the game.

Variation

House Rule: Would you like a reward for successful fishing? At our house, if you get the card you asked for, either from the other player or from the fishing pond, you get a free turn and may ask any player for another card.

Notes & House Rules

Shut the Box

MATH CONCEPTS: addition, number bonds up to nine.
PLAYERS: two or more.
EQUIPMENT: paper and pencil or pen for each player, two six-sided dice.

Set-Up

Players make their own gameboards by writing the numbers 1–9 on a piece of blank paper, or 1–12 for older players. These may be decorated and laminated for frequent play, in which case each player will need an erasable marker or nine tokens for covering the numbers.

Or make a fancy gameboard with flaps using the designs on the next two pages. Decorate as desired.

How to Play

On your turn, roll the dice and add the numbers together. Cross out or cover one or more numbers that add up to make that sum. For instance, if you roll a six and a four, you could cover the 8 and 2, or the 6 and 4, or the 7, 2, and 1, or any other combination that makes a sum of ten. If you cannot cover the full amount of your roll, you don't get to cover anything. Pass the dice to the next player.

If all the higher numbers are already covered, a player may choose to roll only one die. The first player to cross out or cover all the numbers wins.

You can use playing cards as your "gameboard."
Start with all the cards in a suit face up.
How many can you turn down?

Cut only this far.

Fold gameboard here.

Copy gameboard on card stock or regular paper.

- Cut away excess margins.
- Cut along each solid line above, from the outer edge of the paper to the dotted line.
- Fold along the dashed center line.
- Put two or three staples in the uncut area to hold the gameboard together.

Begin playing with all flaps folded open.

1 2 3 4 5 6 7 8 9

Cut only this far.

Fold gameboard here.

Copy gameboard on card stock or regular paper.
- Cut away excess margins.
- Cut along each solid line above, from the outer edge of the paper to the dotted line.
- Fold along the dashed center line.
- Put two or three staples in the uncut area to hold the gameboard together.

Begin playing with all flaps folded open.

1 2 3 4 5 6 7 8 9 10 11 12

Notes & House Rules

Shut the Box Variations

Score the Box

Traditionally, each player takes a single, long turn. You keep playing until you roll a sum you can't cover, and then you add up the uncovered numbers to make your score. Pass the dice to the next player. Whoever gets the lowest score wins the game.

Shut the Box Solitaire

Play by yourself, throwing the dice and covering numbers. If you make all the numbers, you win. But if you roll a throw that you can't play, that's the end of the game.

Plus or Minus

Players may choose to cover either the sum or difference of the numbers on their dice. Subtraction offers more options near the end of the game, when there are only a few numbers left to cover.

Phone Number Cover-Up

Make a gameboard with the digits of your phone number (or any other number you want your child to memorize). Decorate as desired and laminate for repeated play.

The Long Game

(For two players.)

The first player tries to cover all the numbers on the gameboard, going until he rolls a number he can't play. If that last roll was a double, he gets to roll once more, and if he can cover that free roll, he keeps going until he's stumped again.

Then the second player takes the same gameboard, but she tries to uncover all the numbers that the first player covered. She keeps rolling until she gets a number she can't uncover (taking a free turn if that roll was a double), then passes the board back to the first player.

Whoever succeeds in covering or uncovering all the numbers wins the game.

Notes & House Rules

I Have, You Need

MATH CONCEPTS: addition, subtraction, number bonds.
PLAYERS: two or more. (a cooperative game).
EQUIPMENT: none.

Set-Up

Number bonds are pairs of numbers that add together to make a target sum. Players agree on the target sum, often a *benchmark number* like 10, 20, 50, 100, and so on.

Also agree on a time limit, such as:

- until everyone has a certain number of turns
- until all the numbers are used (for low target sums like 10 or 20)
- until the car reaches a certain landmark
- until dinner arrives at the restaurant

How To Play

The first player begins by naming any number less than the target: "I have N. You need?"

The second play says the partner that bonds with N to make the target. Then the second player names a new number that hasn't yet been used.

Each player in turn says the bond partner for the previous player's number, then names a new number of their own.

Players win by keeping the game going without a mistake to break the streak.

Example with a Target of 100

"I have 30. You need?"
   "70. I have 45. You need?"
   "55. I have 27. You need?"

Optional: Players Justify Their Answers

"I have 30. You need?"
   "70, because 3 tens plus 7 tens = 100. I have 45. You need?"
   "Well, 45 is less than half of 100, so I need 50 plus a little more. 55. I have 27. You need?"
   "Oh, a tough one! 27 is three less than 30, so I need 70 plus three more. 73. I have…"

For More Information

Nine Cards

This face-up variation of the classic Tens Concentration comes from Constance Kamii's *Young Children Continue to Reinvent Arithmetic, 2nd Grade: Implications of Piaget's Theory, 2nd ed.*, written with Linda Leslie Joseph, Teachers College Press, 2004.

Shut the Box

This traditional English pub game, also known as Canoga, may have been played as early as the twelfth century in Normandy. The "box" is a wooden tray with a row of numbers one to nine along the top length, each with a cover that can either slide or swing to hide the number. As a gambling game, each player would ante an agreed amount into a pool, which would be awarded to the winner.

Phone Number Cover-Up

First grade teacher Sharon McGlohn created this game, which was shared by Alice P. Wakefield in *Early Childhood Number Games: Teachers Reinvent Math Instruction*, Allyn & Bacon, 1998.

I Have, You Need

Many teachers have played cooperative number bond games with their students over the years. I believe the name "I Have, You Need" was coined by Kim Montague, who cohosts the *Math is Figure-Out-Able* podcast with Pam Harris.

♦ podcast.mathisfigureoutable.com

♦ ♦ ♦

"When children are given worksheets, the teacher makes all the decisions about what to do and which answer is correct. When they play math games, by contrast, they can learn to make decisions for themselves about what is fair, which answer is correct, and whether or not it makes sense to change a rule."

—Constance Kamii and Linda Leslie Joseph,
Young Children Continue to Reinvent Arithmetic

TABLETOP MATH GAMES COLLECTION

Games with Numbers to 100

6 Ways to Play Math with Primary Students

Notes & House Rules

Thirty-One

MATH CONCEPTS: addition to thirty-one, thinking ahead.
PLAYERS: best for two.
EQUIPMENT: one deck of playing cards.

How to Play

Lay out the ace to six of each suit in a row, face up and not overlapping, one suit above another. You will have one column of four aces, a column of four twos, and so on—six columns in all.

The first player flips a card upside down and says its number value. Players alternate, each time turning down one card, mentally adding its value to the running total, and saying the new sum out loud. The player who exactly reaches thirty-one, or who forces the next player to go over that sum, wins the game.

Variation

For a shorter game, use only the ace to four of each suit. Play to a target sum of twenty-two.

Ready to start a game of Thirty-One.

Notes & House Rules

Push the Penny

MATH CONCEPTS: addition to one hundred, thinking ahead.
PLAYERS: two or more.
EQUIPMENT: your choice of hundred chart (or 0–99 chart), one deck of playing cards (face cards removed), and a penny or other small token.

How to Play

Place the chart in the middle of the table, with the penny on the starting number.

Deal three cards per player and set the rest of the deck face down as a draw pile.

On your turn, lay one card from your hand face up on the discard pile. Add your number to the current total and move the penny forward to that new sum. Then draw to replenish your hand.

If adding the number on your card would move the penny past the end of the chart, then you must subtract your number and move backward instead. Whoever pushes the penny to the final square by exact count wins the game.

Variations

Play two cards from your hand and choose whether you want to add or subtract each of those numbers to find out where the penny moves.

Or start at the greatest number on your chart and subtract every time, pushing the penny down to zero.

Many people find the bottoms-up hundred chart more logical than the traditional top-down version.

It makes intuitive sense to have the numbers grow as they climb up the page.

91	92	93	94	95	96	97	98	99	100
81	82	83	84	85	86	87	88	89	90
71	72	73	74	75	76	77	78	79	80
61	62	63	64	65	66	67	68	69	70
51	52	53	54	55	56	57	58	59	60
41	42	43	44	45	46	47	48	49	50
31	32	33	34	35	36	37	38	39	40
21	22	23	24	25	26	27	28	29	30
11	12	13	14	15	16	17	18	19	20
1	2	3	4	5	6	7	8	9	10

1	2	3	4	5	6	7	8	9	10
11	12	13	14	15	16	17	18	19	20
21	22	23	24	25	26	27	28	29	30
31	32	33	34	35	36	37	38	39	40
41	42	43	44	45	46	47	48	49	50
51	52	53	54	55	56	57	58	59	60
61	62	63	64	65	66	67	68	69	70
71	72	73	74	75	76	77	78	79	80
81	82	83	84	85	86	87	88	89	90
91	92	93	94	95	96	97	98	99	100

91	92	93	94	95	96	97	98	99	100
81	82	83	84	85	86	87	88	89	90
71	72	73	74	75	76	77	78	79	80
61	62	63	64	65	66	67	68	69	70
51	52	53	54	55	56	57	58	59	60
41	42	43	44	45	46	47	48	49	50
31	32	33	34	35	36	37	38	39	40
21	22	23	24	25	26	27	28	29	30
11	12	13	14	15	16	17	18	19	20
1	2	3	4	5	6	7	8	9	10

0	1	2	3	4	5	6	7	8	9
10	11	12	13	14	15	16	17	18	19
20	21	22	23	24	25	26	27	28	29
30	31	32	33	34	35	36	37	38	39
40	41	42	43	44	45	46	47	48	49
50	51	52	53	54	55	56	57	58	59
60	61	62	63	64	65	66	67	68	69
70	71	72	73	74	75	76	77	78	79
80	81	82	83	84	85	86	87	88	89
90	91	92	93	94	95	96	97	98	99

90	91	92	93	94	95	96	97	98	99
80	81	82	83	84	85	86	87	88	89
70	71	72	73	74	75	76	77	78	79
60	61	62	63	64	65	66	67	68	69
50	51	52	53	54	55	56	57	58	59
40	41	42	43	44	45	46	47	48	49
30	31	32	33	34	35	36	37	38	39
20	21	22	23	24	25	26	27	28	29
10	11	12	13	14	15	16	17	18	19
0	1	2	3	4	5	6	7	8	9

Notes & House Rules

Dollar Nim

MATH CONCEPTS: subtraction within one hundred, value of coins, thinking ahead.
PLAYERS: best for two.
EQUIPMENT: none.

How to Play

This is a mental math game, designed to be played in the car. Start by imagining a pile of money equal to $1, or 100 cents. On your turn, "remove" any coin you like—quarter, dime, nickel, or penny. Say which coin you are taking and the new value of the pile. The player who claims the last coin wins the game.

Variations

Allow half-dollar coins, if you wish, but no dollar coins—unless you are trying to demonstrate what mathematicians mean by a "trivial" problem.

HUNDRED CHART NIM: If players have trouble doing the subtraction mentally, you may use a penny or other small token to keep track of your value on a printed hundred chart. Whoever wins gets to keep the penny.

MAKING CHANGE: Play with real coins. Each player starts with one quarter, two dimes, three nickels, and four pennies. Or pick different coins, as long as everyone starts with the same collection.

On your turn, discard one of your coins to a pile on the table. If there are any smaller coins in the pile, you may take back change up to one cent less than the value of the coin you put in. The last player who has money wins the game.

ALIEN MONEY: Imagine an alien civilization. What sorts of coins would they use? Perhaps the creatures have three fingers on each hand, so their coins are all multiples of three.

Or maybe the alien society loves math so much that they put mathematicians on their coins instead of political leaders, and their coins are based on prime numbers.

What is the value of an alien dollar? How would your aliens play Dollar Nim?

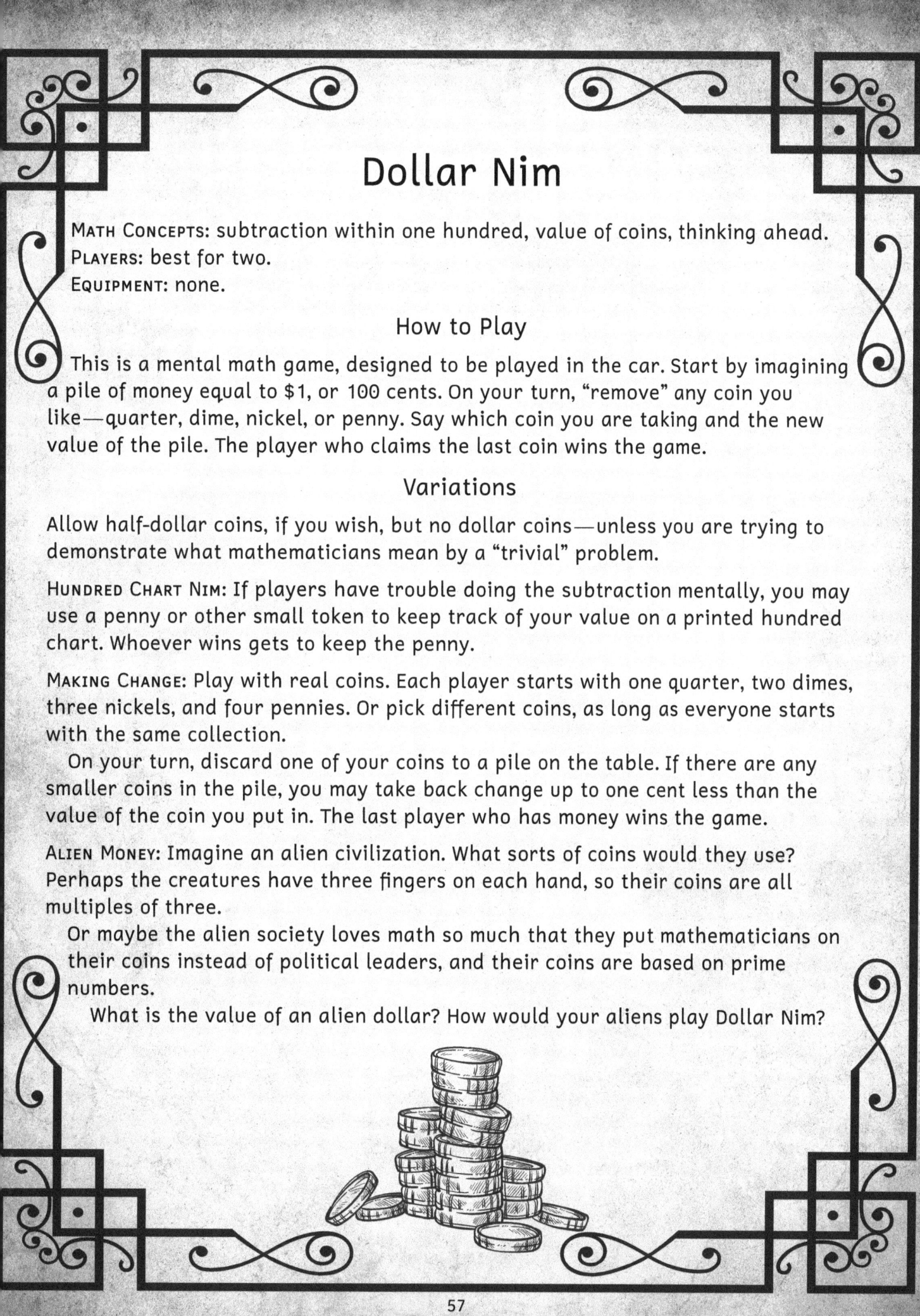

Notes & House Rules

Countdown

MATH CONCEPTS: subtraction within one hundred, mental math, thinking ahead.
PLAYERS: any number.
EQUIPMENT: none.

How to Play

Agree on a starting number, such as one hundred.

The first player subtracts any amount less than the starting number, saying aloud how much was subtracted and the amount that remains.

On each succeeding turn, players subtract any amount from one up to twice as much as the previous move—but keep in mind that your opponent will then be able to subtract up to twice as much as you do.

The player who gets to zero wins the game.

Variations

Keep track of the current number by moving a penny or other token on your choice of hundred or ninety-nine chart.

FIBONACCI NIM: Play as a strategy game for two players.

MULTI-PILE FIBONACCI NIM: Using coins or other tokens, create two or more piles of varying sizes. The first player takes any number of coins from a single pile.

On each succeeding turn, players may take up to twice as many as the previous turn, but only from a single pile. The player who takes the last coin wins the game.

Notes & House Rules

Euclid's Game

MATH CONCEPTS: subtraction within one hundred, number patterns.
PLAYERS: two or more.
EQUIPMENT: blank paper and pen or pencil, or printed hundred chart and a highlighter to mark numbers.

How to Play

Allow the youngest player choice of moving first or second; in future games, allow the loser of the last game to choose.

The first player picks a number from one to one hundred and writes that number on the paper (or marks that square on the hundred chart). The second player writes or marks any other number, except that the second number may not be exactly double or exactly half the first choice.

On each succeeding turn, players subtract any two marked numbers to find and write a difference that has not yet been taken. Play alternates until no more numbers can be made.

The player who marks the last number wins the game.

For Advanced Players

Play several rounds of Euclid's Game on printed hundred charts. Circle the original pair of numbers in each game, and use a highlighter to mark all the numbers you use. Then study your collection of finished gameboards.

- ♦ Do you notice any patterns in the numbers marked on each gameboard?
- ♦ Can you explain why some games have few numbers marked while others have many?
- ♦ If you knew the first two numbers, would you be able to predict how many squares would be marked in the end?

Notes & House Rules

Number Grid Tic-Tac-Toe

MATH CONCEPTS: place value, number patterns, thinking ahead.
PLAYERS: two or more.
EQUIPMENT: blank hundred chart or grid paper with squares large enough to write in, colored markers or pencils.

Set-Up

If using grid paper, mark ten columns as your gameboard. Shade in the other squares as out of bounds.

Write the number one in the first square, in black (or plain pencil) to mark it as a "wild" space that can be used by any player. Also write one hundred in the last square, or whatever number would be the greatest on your grid paper.

Each player will need a different color marker, pen, or pencil to distinguish their numbers from all the others.

How To Play

On your turn, choose a blank square and write in the number that belongs in that square according to the standard hundred-chart counting pattern: numbers increase by one in each square moving across the row and continue counting in each successive row down the page.

With two players, whenever you get four number squares in a row, either vertically, horizontally, or diagonally, connect those squares and score a point. The same number may be used on more than one row, as long as the rows go in different directions.

With more than two players, score a point when you get three in a row.

Play until all the squares are filled or until it's impossible for any player to make another row. Whoever scores the most points wins the game.

Variations

NUMBER GRID CHALLENGE: Instead of marking the number one in the first square, begin with any number you like.

- For an easier challenge, choose a number that ends in one, like 371, so the hundred chart rows and columns still follow the familiar pattern.
- For a harder challenge, pick any random number.
- You can even count by fractions (¼, ²⁄₄, ¾, 1, 1 ¼, etc.) or by decimal fractions (0.1, 0.2, 0.3, etc.) if you choose.

1									
									100

1
120

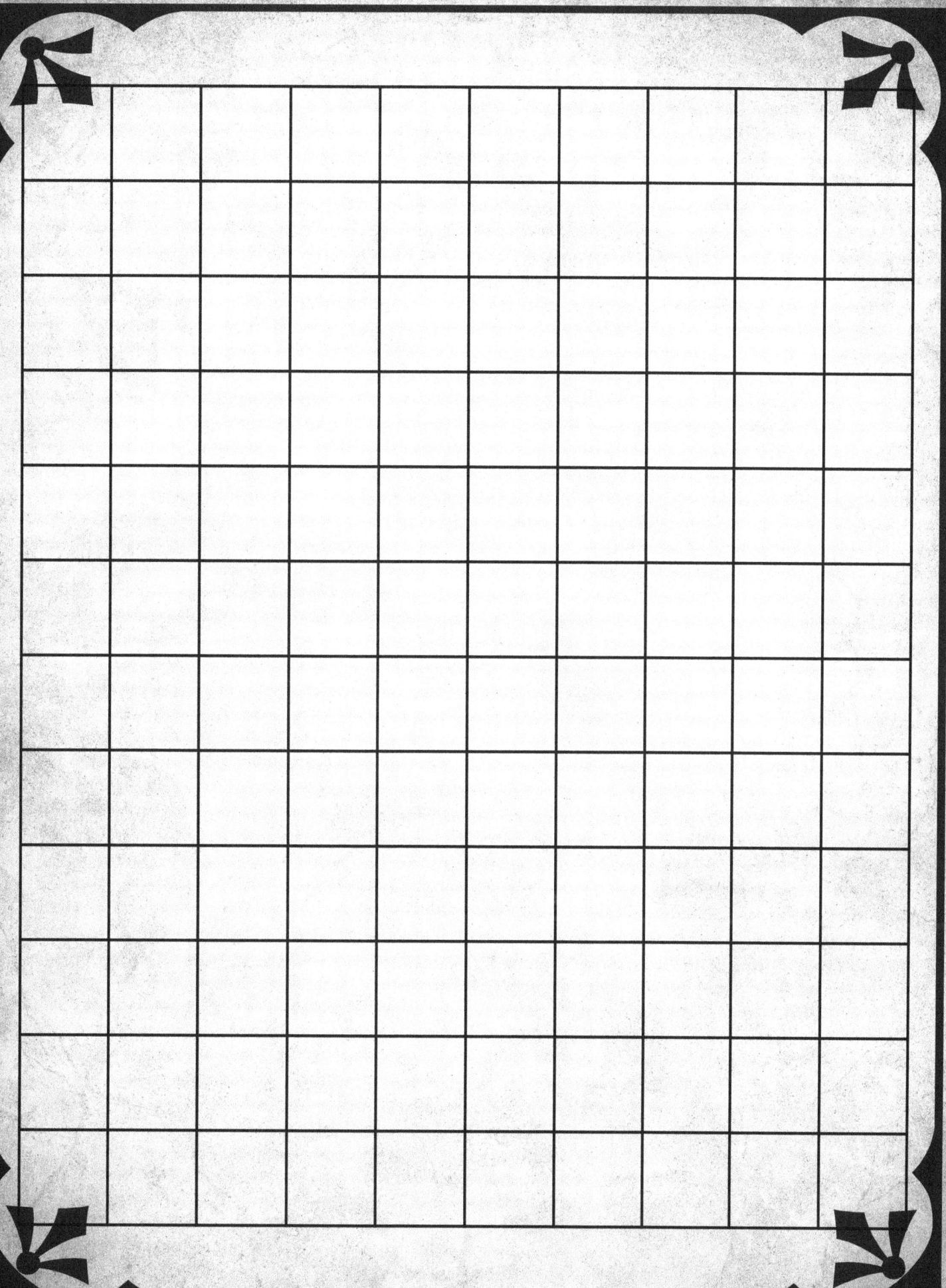

For More Information

Thirty-One (and a Puzzle)

Thirty-One comes from British mathematician Henry Dudeney's classic book, *The Canterbury Puzzles*, published in 1907.

Push the Penny

This is a modification of the 99 Game from Julie Reulbach's collection "Math Games for Classrooms."

- ispeakmath.org/2013/11/18/math-games-collection-on-google-docs-add-your-game-today

Dollar Nim

Nadine Block invented Dollar Nim on the way home from a camping trip, and her husband Patrick Vennebush shared it on his blog.

- mathjokes4mathyfolks.wordpress.com/2011/06/27/dollar-nim

Making Change is based on James Ernest's game Fight, later renamed Pennywise. I found it at John Golden's Math Hombre blog.

- mathhombre.blogspot.com/2009/08/money-games.html

Countdown

Countdown can also be called Fibonacci Nim. It was originally described by M. J. Whinihan and is included in the book *Winning Ways for Your Mathematical Plays, Volume 4*, by Elwyn R. Berlekamp, John H. Conway, and Richard K. Guy. I discovered it in an online lesson by Geoff Patterson.

Euclid's Game

A. J. Cole and A. J. T. Davie invented the Euclid game and analyzed it for an article in *The Mathematical Gazette*, December 1969.

Number Grid Tic-Tac-Toe

This game was invented by math teacher Joe Schwartz and shared on his blog Exit 10A.

- exit10a.blogspot.com/2016/01/i-like-this-game-because-you-have-to.html

Yan Tan Tethera

A COOPERATIVE GAME
OF MENTAL MATH
AND COUNTING SHEEP

Notes & House Rules

The Joyful Challenge of Mental Exercise

Throughout the centuries, amateur mathematicians and professionals alike have enjoyed tinkering with recreational mathematics—math we do just for the fun of it, testing our wits against a puzzle, enjoying the challenge.

Math professor David Singmaster gives an example in his essay, "The Unreasonable Utility of Recreational Mathematics":

"Mathematical recreations are as old as mathematics itself. The earliest piece of Egyptian mathematics, the Rhind Papyrus, has a problem where there are 7 houses, each house has 7 cats, each cat ate 7 mice, each mouse would have eaten 7 ears of spelt and each ear of spelt would produce 7 hekat of spelt. A similar problem of adding powers of 7 occurs in Fibonacci, in a few later medieval texts and in the children's riddle rhyme 'As I was going to St. Ives.'

"There is no greater learning experience than trying to solve a good problem. Recreational mathematics provides many such problems and almost every problem can be extended or amended. Hence recreational mathematics is also a treasury of problems for student investigations.

"Because of its long history, recreational mathematics is an ideal vehicle for communicating historical and multicultural aspects of mathematics."

One of my favorite recreational math puzzles is *number yoga*, where we choose a set of numbers and combine them with arithmetic operations to create expressions for different numbers.

The classic number yoga puzzle is Four Fours:

♦ Using exactly four copies of the digit 4, along with whatever arithmetic operations you know, create expressions for all the numbers from one to infinity (or as high as you can go).

Math teachers like to challenge their students with the Year Game:

♦ Using each of the four digits in this year exactly once, along with whatever arithmetic operations you know, create expressions for all the numbers from one to one hundred (or as high as you can go).

Yan Tan Tethera

Math Concepts: addition, subtraction, multiplication, division, fractions, decimals, exponents, factorials, order of operations, multistep mental math.
Players: any number (a cooperative game).
Equipment: gameboard or paper, dice or playing cards, pen or pencil.

History

Yan Tan Tethera is a simpler version of number yoga, a cooperative game to stretch and work out mental muscles, based on the traditional British Celtic sheep-counting system used in northern England.

As sheep walked through a gate or other narrow space, the shepherd counted up to *jigget* (twenty), picked up a small rock as a tally marker, and then started again at *yan* (one).

Base Twenty

In our base-ten number system, the columns are powers of ten:

10 × 10 = hundreds	tens	ones	.	tenths

British Celts used a base-20 number system, counting powers of twenty:

20 × 20 = four hundreds	twenties	ones	.	twentieths

But the yan-tan-tethera counting system didn't use columns. How did rocks help the shepherd know how many sheep passed the gate?

How To Play

Roll four dice or draw four cards to generate random numbers. With playing cards, you can remove the face cards or use Jack = 11, Queen = 12, and King = 13. Leave the dice or cards on the table where everyone can see them, or write them into the squares on the gameboard.

The first player combines one or more of the random numbers to create an arithmetic expression equal to *yan* (one). Then each player in turn creates the next counting-sheep number in order.

You win if you find an expression for all the sheep-counting numbers from *yan* to *jiggit*. This game has an element of luck. Some sets of numbers work all the way to *jiggit*, while other sets will stump even the sharpest players.

Creating Number Expressions

Players may add, subtract, multiply, or divide the numbers or use any other arithmetic operations they know. For example, a set of 3, 4, 7, and 9 could make:

$$Yan = 4 - 3$$
$$Tan = 9 - 7$$
$$Tethera = 3$$
but not
$$Methera = 3 + 7/7,$$
because you only have one digit 7.

Consider using fractions, decimals, exponents, or factorials to find the toughest numbers. Factorials are written with an exclamation point:

$$2! = 2 \times 1 = 2$$
$$3! = 3 \times 2 \times 1 = 6$$
$$4! = 4 \times 3 \times 2 \times 1 = 24$$
$$5! = 5 \times 4 \times 3 \times 2 \times 1 = 125, \text{etc.}$$

Variations

Allow players to pass when they can't think of an expression for the current number. On future turns, players may solve either the next number in the counting sequence or any number that has been skipped.

To challenge older players, don't allow single-digit expressions.

MAYBE-FOUR MAYBE-FOURS: On each turn, you must choose only one of your random numbers, but you can use as many copies of it as you wish.

Sample Game

Use one or more of the numbers in the boxes, along with any math you know, to create the sheep-counting number series. Start at *yan*, and then each player in turn makes *tan*, *tethera*, *methera*, and so on. How high can you count?

$$\boxed{4} \quad \boxed{4} \quad \boxed{4} \quad \boxed{4}$$

Yan = 1 = $44/44$	Yanadix = 11 =
Tan = 2 = $4/4 + 4/4$	Tanadix = 12 =
Tethera = 3 = $4 - 4/4$	Tetheradix = 13 = $4! - 44/4$
Methera = 4 = (...and so on.)	Metheradix = 14 = $4 \times (4 - .4) - .4$
Pip = 5 = (I'll do the toughest ones.)	Bumfit = 15 =
Tayter = 6 =	Yanabum = 16 =
Layter = 7 =	Tanabum = 17 =
Overa = 8 =	Tetherabum = 18 = $44 \times .4 + .4$
Covera = 9 =	Metherabum = 19 = $4! - (4 + 4/4)$
Dix = 10 =	Jiggit = 20 =

Yan Tan Tethera Gameboard

Use one or more of the numbers in the boxes, along with any math you know, to create the sheep-counting number series. Start at *yan*, and then each player in turn makes *tan*, *tethera*, *methera*, and so on. How high can you count?

| 1 | 2 | 3 | 4 |

Yan = 1 =	Yanadix = 11 =
Tan = 2 =	Tanadix = 12 =
Tethera = 3 =	Tetheradix = 13 =
Methera = 4 =	Metheradix = 14 =
Pip = 5 =	Bumfit = 15 =
Tayter = 6 =	Yanabum = 16 =
Layter = 7 =	Tanabum = 17 =
Overa = 8 =	Tetherabum = 18 =
Covera = 9 =	Metherabum = 19 =
Dix = 10 =	Jiggit = 20 =

Yan Tan Tethera

Use one or more of the numbers in the boxes, along with any math you know, to create the sheep-counting number series. Start at *yan*, and then each player in turn makes *tan, tethera, methera,* and so on. How high can you count?

| 1 | 3 | 6 | 10 |

Yan = 1 =	Yanadix = 11 =
Tan = 2 =	Tanadix = 12 =
Tethera = 3 =	Tetheradix = 13 =
Methera = 4 =	Metheradix = 14 =
Pip = 5 =	Bumfit = 15 =
Tayter = 6 =	Yanabum = 16 =
Layter = 7 =	Tanabum = 17 =
Overa = 8 =	Tetherabum = 18 =
Covera = 9 =	Metherabum = 19 =
Dix = 10 =	Jiggit = 20 =

Yan Tan Tethera

Use the numbers in the boxes along with any math you know to create the sheep-counting series. Start at *yan*, and then each player in turn makes *tan*, *tethera*, *methera*... How high can you count?

3   6   7   12

Yan = 1 =	Yanadix = 11 =
Tan = 2 =	Tanadix = 12 =
Tethera = 3 =	Tetheradix = 13 =
Methera = 4 =	Metheradix = 14 =
Pip = 5 =	Bumfit = 15 =
Tayter = 6 =	Yanabum = 16 =
Layter = 7 =	Tanabum = 17 =
Overa = 8 =	Tetherabum = 18 =
Covera = 9 =	Metherabum = 19 =
Dix = 10 =	Jiggit = 20 =

Yan Tan Tethera

Use playing cards or dice to generate four random numbers. Use these numbers along with any math you know to create the sheep-counting series. Start at *yan*, and each player in turn makes *tan*, *tethera*, *methera*, and so on. How high can you count?

Yan = 1 =	Yanadix = 11 =
Tan = 2 =	Tanadix = 12 =
Tethera = 3 =	Tetheradix = 13 =
Methera = 4 =	Metheradix = 14 =
Pip = 5 =	Bumfit = 15 =
Tayter = 6 =	Yanabum = 16 =
Layter = 7 =	Tanabum = 17 =
Overa = 8 =	Tetherabum = 18 =
Covera = 9 =	Metherabum = 19 =
Dix = 10 =	Jiggit = 20 =

Yan Tan Tethera

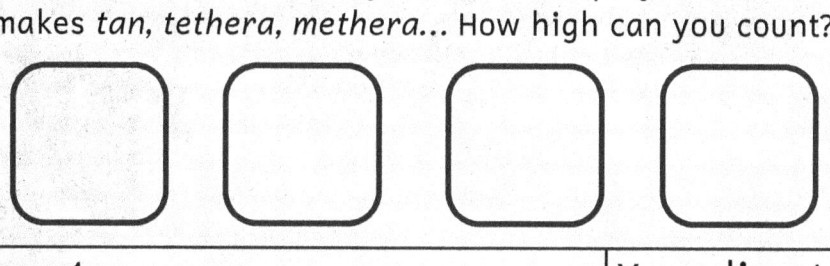

Use playing cards or dice to generate four random numbers. Use any math you know to create the sheep-counting series. Start at *yan*, and each player in turn makes *tan, tethera, methera...* How high can you count?

Yan = 1 =	Yanadix = 11 =
Tan = 2 =	Tanadix = 12 =
Tethera = 3 =	Tetheradix = 13 =
Methera = 4 =	Metheradix = 14 =
Pip = 5 =	Bumfit = 15 =
Tayter = 6 =	Yanabum = 16 =
Layter = 7 =	Tanabum = 17 =
Overa = 8 =	Tetherabum = 18 =
Covera = 9 =	Metherabum = 19 =
Dix = 10 =	Jiggit = 20 =

For More Information

Recreational Mathematics

You can read David Singmaster's essay, "The Unreasonable Utility of Recreational Mathematics" on the Internet Archive Wayback Machine:

- web.archive.org/web/20020207051215/http://anduin.eldar.org/~problemi/singmast/ecmutil.html

Number Yoga

The term "number yoga" was coined by Bill Lombard and Brad Fulton for the puzzles in a handout of a session given at CMC-N (Asilomar) Dec 2009: "Games and Puzzles that Reach the Kids and Teach the Standards."

Lombard's excellent blog has disappeared from the internet, but you can find his workshop notes on the Wayback Machine:

- web.archive.org/web/20120916085011/http://www.mrlsmath.com/wp-content/uploads/2008/10/Games-and-Puzzles-that-Reach-the-Kids-and-Teach-the-Standards-download-file.pdf

Four Fours

Pat Ballew traces the history of the Four Fours number puzzle, and David Wheeler puts together "The Definitive Four Fours Answer Key":

- pballew.blogspot.com/2024/02/before-there-were-four-fours-there-were.html
- dwheeler.com/fourfours

◆ ◆ ◆

"Puzzles and Games are a great way to build enthusiasm, excitement, and skills in a math classroom. When these are used properly, there is always an underlying mathematical theme that is being explored or reinforced. Students learn the value of logical thinking, proper planning, and long-term focus.

"Opportunities for transitioning from number sense to algebra thinking abound with puzzles. Number properties show their strength in supporting algebra throughout puzzle activities. It's always a pleasure to watch the enthusiasm and skills grow in students!"

—BILL LOMBARD, *PUZZLES AND GAMES*

TABLETOP MATH GAMES COLLECTION

Games for Modeling Multiplication

Build Visualization Skills to Deepen Understanding

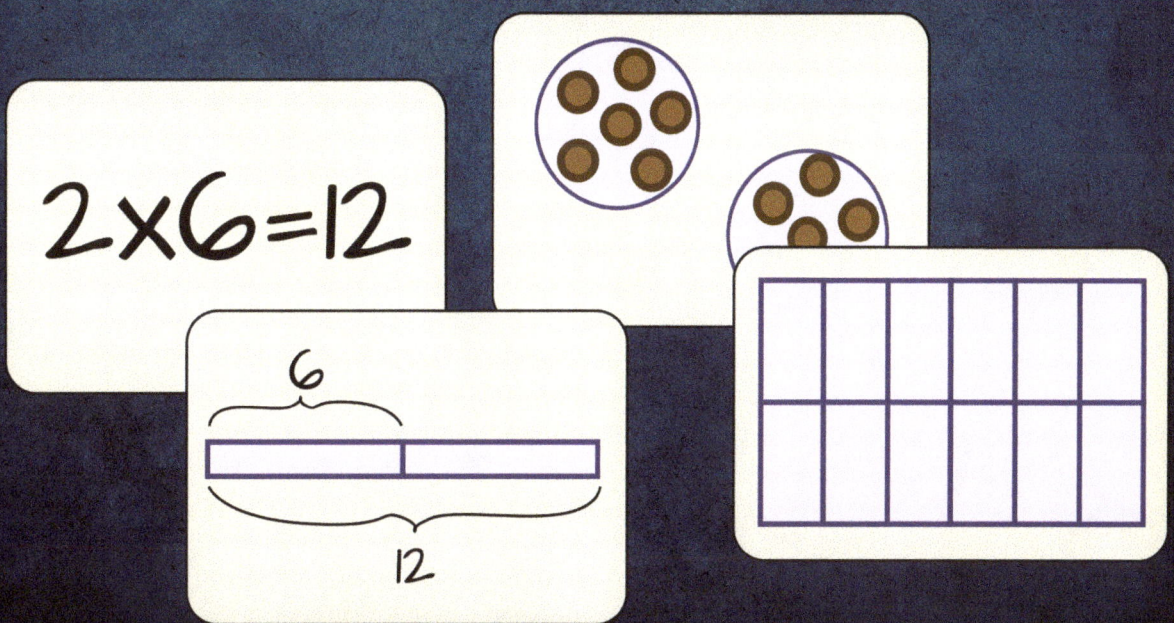

Notes & House Rules

What Is a Math Model?

Teachers and parents know that every elementary math student needs to master the number facts, the basic relationships between all the one-digit numbers. To memorize so many details can seem like an unending task.

Too often, we adults are tempted to stress the rote aspect of such memory work, which makes our children lose their focus on what the numbers mean. But if we concentrate first on learning and using *math models*—physical or pictorial representations that help students make sense of mathematical concepts—we give our kids a strong foundation for middle school math.

Models give us a way to form and manipulate an image of an abstract concept, such as multiplication. When teaching young students, we act out ideas using blocks, cookies, or pieces of construction paper. Older students firm up their understanding by drawing pictures. As kids grow up and face more abstract, numbers-only problems, these pictures remain in their minds, an always-ready tool to help them reason their way through multiplication problems.

The reason we teach more than one model is that none of them can fit every type of problem. Models are thinking tools, and every tool has its limit.

Keep in mind semantics expert Samuel Hayakawa's first principle: "The symbol is *not* the thing symbolized; the word is *not* the thing; the map is *not* the territory it stands for."

And the math model is not the number itself.

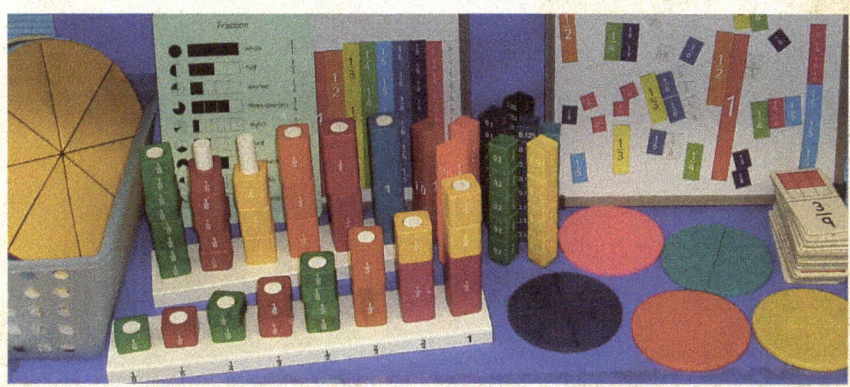

Your math curriculum may offer a variety of physical or pictorial models to help build comprehension.

Photo by misskprimary flickr.com/photos/misskprimary/1037165017 (CC-BY-2.0).

Multiplication Model Cards

You can work together to create a deck of math model cards, sketching the pictures on index cards or on a stack of old business cards with blank backs. A deck of math model cards should include 10–15 sets, or *books*. Each book consists of four cards—the multiplication equation and the following three pictures:

(1) Set Model: "___ sets of ___ objects per set"

This model represents discrete (countable) items collected into groups: apples per basket, pennies per dime, or cookies per child. The set model connects the concepts of multiplication and addition, so it is the most common way of introducing multiplication in elementary school textbooks.

(2) Bar (Measurement) Model: "___ units of ___ parts per unit"

This model represents continuous quantities measured out in parts: inches per foot, cups per recipe, dollars per pound of produce, or spaces per jump on a number line. The measurement model can also include scaling, stretching, or shrinking something from its original size, which makes it useful when thinking about fractions.

(3) Rectangular Model: "___ rows of ___ items per row"

In early elementary math, this model represents an array of discrete items: chairs per row, buttons per column, or band members on parade. As students grow, however, the model expands to include continuous rectangular area. At its most mature, this model becomes the basis for many topics in high school math and beyond, including integral calculus. Because of its flexibility, the rectangular model is the most important one for students to master.

Take Your Time

Try making just two or three books each day, while talking about real-life situations the models might symbolize. When you think the deck is finished, lay the cards out on the table in sets, to make sure each book has all its members. Your deck need not include every math fact in the times tables, but should have enough variety to cement the most common multiplication models in your mind.

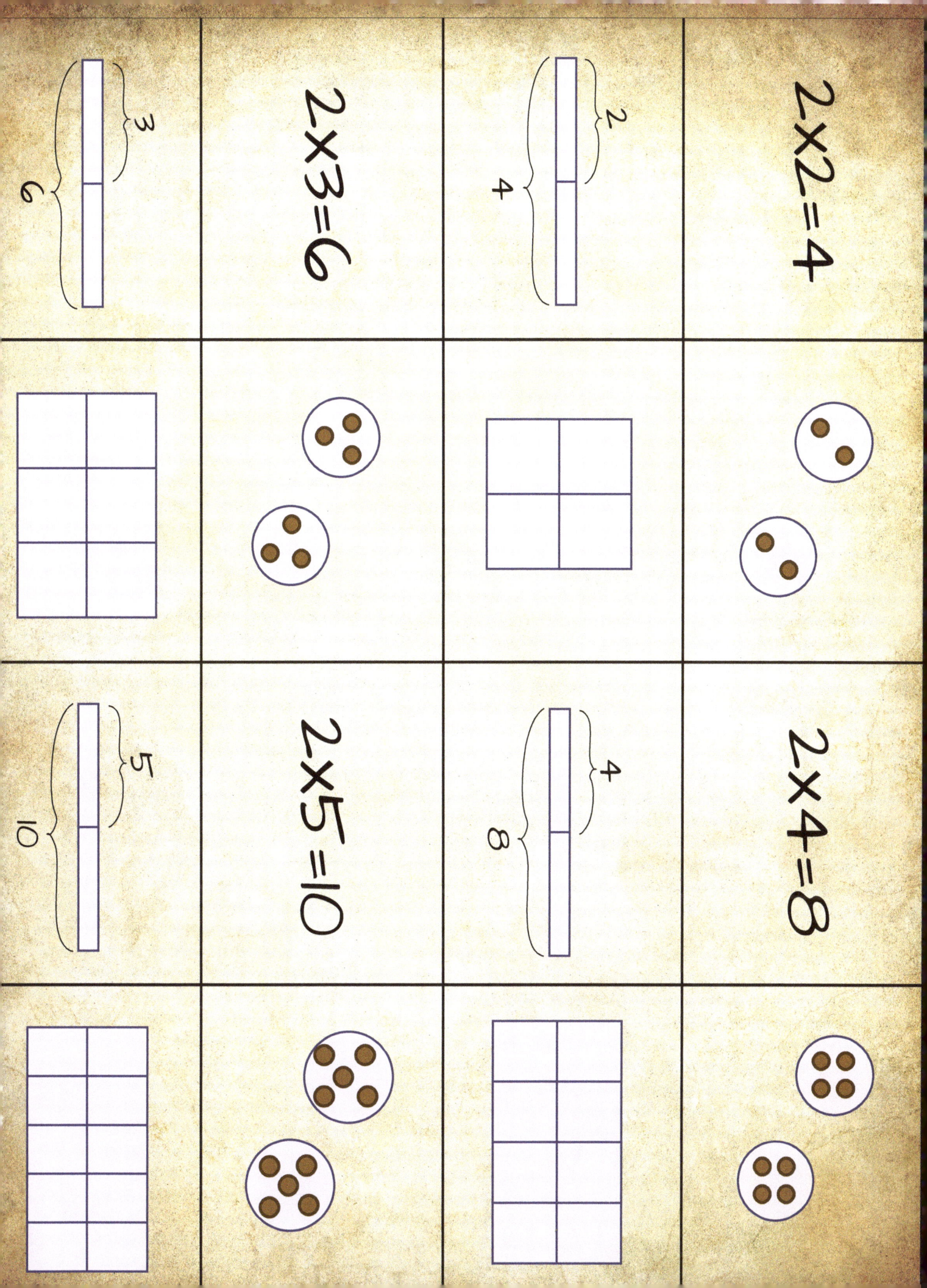

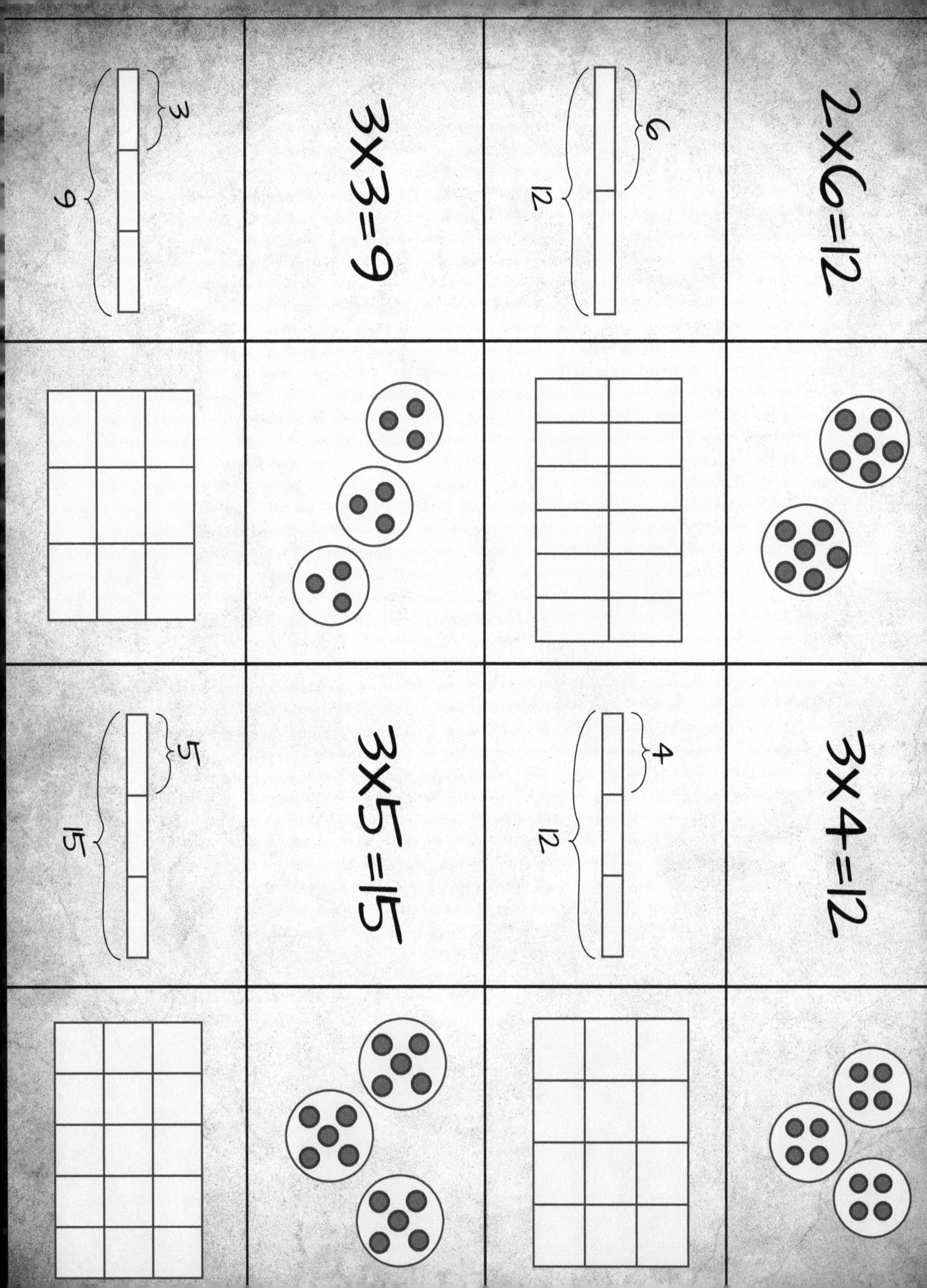

Look for the Ratio

Notice that in each model, the two numbers of a multiplication problem have different roles. One number is *a scale factor,* also called the *multiplier,* which tells you how many sets, units, or rows you are talking about. The other number is *a this-per-that ratio,* also called the *multiplicand.*

In addition and subtraction, numbers count how much stuff you have. If you get more stuff, the numbers get bigger. If you lose some of the stuff, the numbers get smaller. Numbers measure the amount of cookies, horses, dollars, gasoline, or whatever.

The multiplicand ratio is not a counting number, but something new. Something alien, completely abstract. It doesn't count the number of dollars or measure the volume of gasoline but tells the relationship between them, the dollars per gallon, *which stays the same whether you buy a lot or a little.*

A ratio is a relationship number.

This is why you may hear mathematicians say, "Multiplication is *not* repeated addition."

We can use addition to solve whole-number multiplication puzzles, but that will not get us very far. Until we wrestle with and come to understand the concept of a multiplicand ratio, we can never master multiplication.

One book of multiplication model cards.

Notes & House Rules

Twelve Cards

MATH CONCEPTS: multiplication models.
PLAYERS: two or more.
EQUIPMENT: one deck of math model cards.

How to Play

The first player shuffles the deck and then turns up the top twelve cards, placing them face up in a 3 × 4 array: three rows with four cards in each row. That player removes any pairs of cards that show the same product or fraction, keeping them in a personal score stash on the table.

After claiming all visible pairs, the first player passes the deck to the next player. Each player in turn deals out enough cards to fill in the empty spots in the array, captures any matching pairs, and passes the deck on.

On rare occasions, you'll get a dead hand where none of the cards you deal out will match. In that case, deal another set of twelve cards on top of the first. Claim the visible pairs, and also claim any bonus pairs that show up as you reveal the cards underneath.

Cards may only be taken in pairs, so if three cards match, you must leave one of them for the next turn. But if all four cards of a set are showing, you may take both pairs.

The game ends when there aren't enough cards left to fill the holes in the array. Whoever has collected the most cards wins the game.

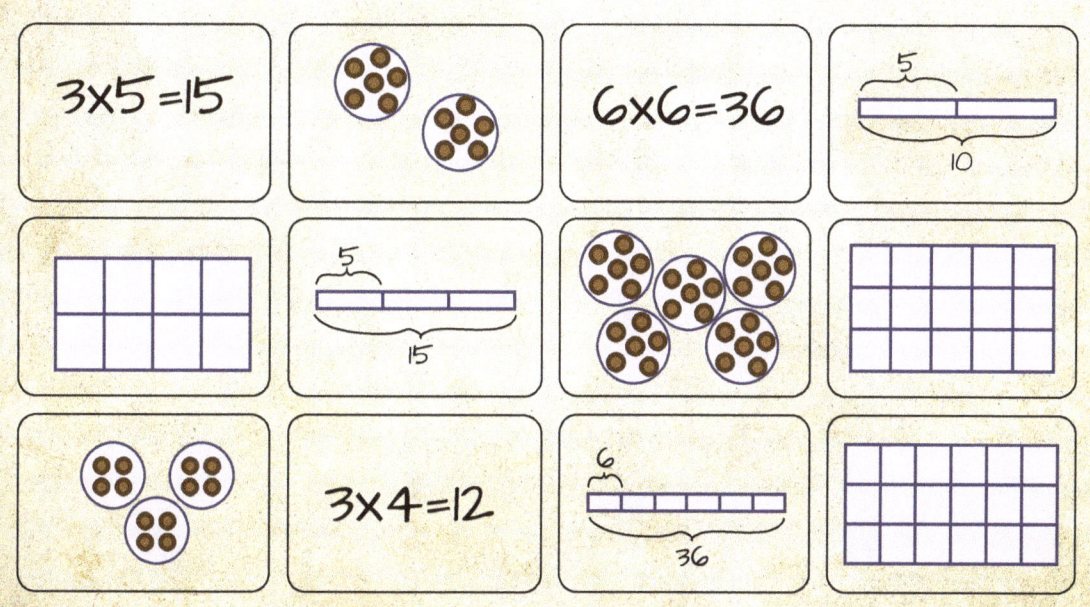

Can you find all four matching pairs of cards? When three cards match, you may choose any two of them and leave the third.

Notes & House Rules

Go Fish

MATH CONCEPTS: multiplication models.
PLAYERS: two or more.
EQUIPMENT: one deck of math model cards, or a double deck for four or more players.

How to Play

Deal seven cards to each player. Players look at their own hands and lay down pairs that show the same product. Each player creates a personal score pile, or fish basket.

Put the remaining cards face down in the center of the table and spread them to make a roughly circular fishing pond.

On your turn, you may ask one other player, "Do you have a _____?" The blank is for the multiplication expression or fraction that matches a card in your hand. For instance, if you have a 3 × 4 card, you might ask, "Do you have a 3 × 4 = 12?" With more than two players, the request must be addressed to a specific person. You may ask, "George, do you have a 2 × 3?" but it is illegal to say, "Does anyone have a 2 × 3?"

- ♦ If George has the card you want, he must give it to you, and you put that card plus its match from your hand into your fish basket.

- ♦ If not, he says, "Go Fish." Then you draw any card you wish from the fishing pond. If you draw a card that matches any card in your hand, add that pair to your basket. Otherwise, add the card to your hand.

If the fishing pond goes dry, every player who has two or more cards in hand places one card face down to replenish the stock. Mix these cards thoroughly before continuing the game.

The game is over when one player runs out of cards. The other players throw their remaining cards into the fishing pond. (Those fish were too small to keep.) Then all players count the cards in their fish basket pile. Whoever caught the most fish wins the game.

Variations

HOUSE RULE: At our house, if you get the card you asked for, either from the other player or from the fishing pond, you get a free turn and may ask any player for another card.

FISH FOR FOUR: Do you enjoy an element of risk? Instead of laying down pairs, players must collect all four cards in a set before adding them to the fish basket. This rule allows players a chance to "steal" what another player asked for in an earlier turn.

Notes & House Rules

Concentration (Memory)

Math Concepts: multiplication models, visual/spatial memory.
Players: any number.
Equipment: one deck of math model cards.

How to Play

Shuffle the cards and lay them all face down on the table, spread out in a single layer. The cards may be placed in an array or arranged in a haphazard cloud, as long as no card covers any other card.

On your turn, flip two cards face up. If the cards match, representing the same product, then you get to take the pair. If they do not match, leave the cards showing for a few seconds so all players can see what they are. Then turn them face down and let the next player take a turn.

Keep the cards you capture in a personal score pile. When all the cards are claimed, whichever player has collected the most is the winner.

Variations

House Rule: How will you handle the frustrating cycle where a player turns up new cards and sees that one of them would match a previously exposed card, but the other player grabs that pair, leaving the first player to try unknown cards again next turn?

At our house, if you find a pair, you get a free turn and can flip over two more cards—which means every player exposes new cards that the next player can use. Free turns expire when there are ten or fewer cards left on the table, to keep one lucky player from claiming all the last pairs.

Mixed Groups: When playing with a wide range of ages, let the younger players flip three cards per turn and keep any two that match.

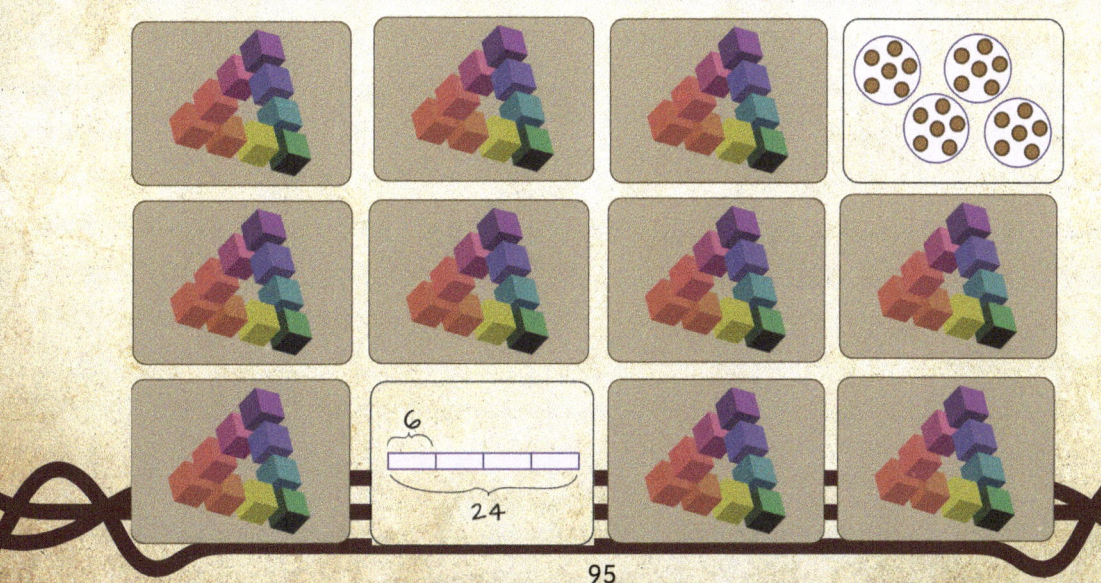

Notes & House Rules

Multiplication Rummy

MATH CONCEPTS: multiplication models.
PLAYERS: two or more.
EQUIPMENT: one deck of math model cards, or a double deck for four or more players.

How to Play

Deal seven cards per player, and place the remaining cards face down as a draw pile. Turn up the top card of the deck to start the discard pile.

On your turn, you may either draw the top card from the deck or pick up the discard pile as far back as desired. But if you pick up more than the top discard, you have to meld the farthest-back card you take.

After drawing, you may *meld*—that is, place three or more matching cards face up on the table in front of you. If you have the fourth card in a set that has already been played (by any player), you may also lay that down in front of you.

Finally, put one card on the discard pile to end your turn. But if you discard a card that could have been played, any other player can call "Rummy!" and meld your discard.

If the deck runs out, take all the discards except the top one. Shuffle these cards and place them face down so the next player has a stack to draw from. Play continues until one player runs out of cards (either by laying them all down or by discarding the last one).

Scoring

Count each player's score as follows:

♦ Each card played on the table is worth 5 points.

♦ For every card remaining in the hand, subtract 2 points.

♦ The player who went out gets a bonus of 15 points.

You may play a single hand, just for fun. Or play several hands, and the first player to reach 300 points wins the game. Or set a different point goal based on how long you want the game to last.

Variation

FLOATER: You must have a discard to end the game, and this must be a card that could not be melded. If you lay down all your cards without a discard, you become a *floater*. Continue to play your turn—drawing or picking from the discard pile and melding cards—until you can go out with a proper discard. (But it is illegal to pick up just the top discard and immediately discard it.)

Notes & House Rules

Multiplication Number Train

MATH CONCEPTS: multiplication models, numerical order, thinking ahead.
PLAYERS: two or more.
EQUIPMENT: one deck of math model cards, or a double deck for four or more players.

Set-Up

Turn all the cards face down on the table and mix them around to make a fishing pond. Each player draws six cards from the pond but does not look at them. Players arrange their cards in a face-down row (train), as shown. When all players are ready, turn the cards in your train face up without changing their position.

Line up the cars of your train.

How to Play

Your goal is to make the products in your train increase from left to right, but of course it will be mixed up to start with.

On your turn, draw one card from the fishing pond.

You have two choices:

♦ Mix the new card back into the fishing pond.

♦ Use the new card to replace one of your others, and then discard the old one.

The first player to complete a train in which the products increase from left to right wins the game.

HOUSE RULE: Decide how strict you will be about the "increases from left to right" rule and repeated numbers. Do matching products count in a valid number train? Or will the player have to keep trying for a card to replace one of the duplicates?

For More Information

Twelve Cards

This face-up variation of the classic Concentration game is adapted from a game in Constance Kamii's *Young Children Continue to Reinvent Arithmetic, 2nd Grade: Implications of Piaget's Theory, 2nd ed.*, written with Linda Leslie Joseph, Teachers College Press, 2004.

Multiplication Rummy

According to the Pagat website, Rummy-style games first appeared in the early twentieth century. Like many card games, Rummy picked up different flavors as it traveled from one player to another over the years. Feel free to modify the rules to fit your family's favorite way to play.

- ♦ pagat.com/rummy/rummy.html

❖ ❖ ❖

"'Mastery' in this context means not just being able to perform calculations with fluency. It is also important to have a good conceptual understanding of numbers, arithmetic, and reasoning, particularly in the context of real-world applications.

"A person can have computational skills without much conceptual understanding. For example, a child can learn the multiplication tables by rote to the point of rapid, automatic recall, without having any understanding of what multiplication is or how it relates to things in the world.

"Math involves a whole lot more than rote learning of a few facts. You can learn to calculate with numbers without any real understanding of the underlying concepts.

"But applying arithmetic to things in the world, to quantities, and understanding the relationships between those quantities, requires considerable understanding of those underlying concepts."

—Keith Devlin, "Wanted: A Mathematical iPod"

TABLETOP MATH GAMES COLLECTION

TIMES TABLE GAMES

7 WAYS TO PLAY MATH WITH PRIMARY STUDENTS

Notes & House Rules

Galactic Conquest

MATH CONCEPTS: multiplication math facts, rectangular area.
PLAYERS: two to four.
EQUIPMENT: graph paper, two six-sided or gaming dice or one deck of playing cards (face cards removed), colored markers.

Set-Up

Players share a single page of graph paper, which represents the galaxy. Each player will need a colored marker to shade in the gameboard squares, and the colors must be different enough to be easily distinguished.

Each player colors one corner square of the grid, as far apart from each other as possible. This is your home star system.

If you are using cards, shuffle the deck and place a draw pile where all players can reach. For beginners, remove the cards for numbers whose multiplication facts they haven't studied.

How to Play

On your turn, roll two dice or draw two cards. Using those numbers as length and width, draw a rectangle that shares at least one side of a grid square with your current territory. If you are using cards, discard face up beside the draw pile.

Your new rectangle may not overlap squares already claimed by any player. Inside this rectangle, write the area (length × width) of your newly conquered space.

The game ends when any player cannot draw a rectangle to match their numbers.

Players add up the areas of all their rectangles, and whoever has conquered the most territory wins.

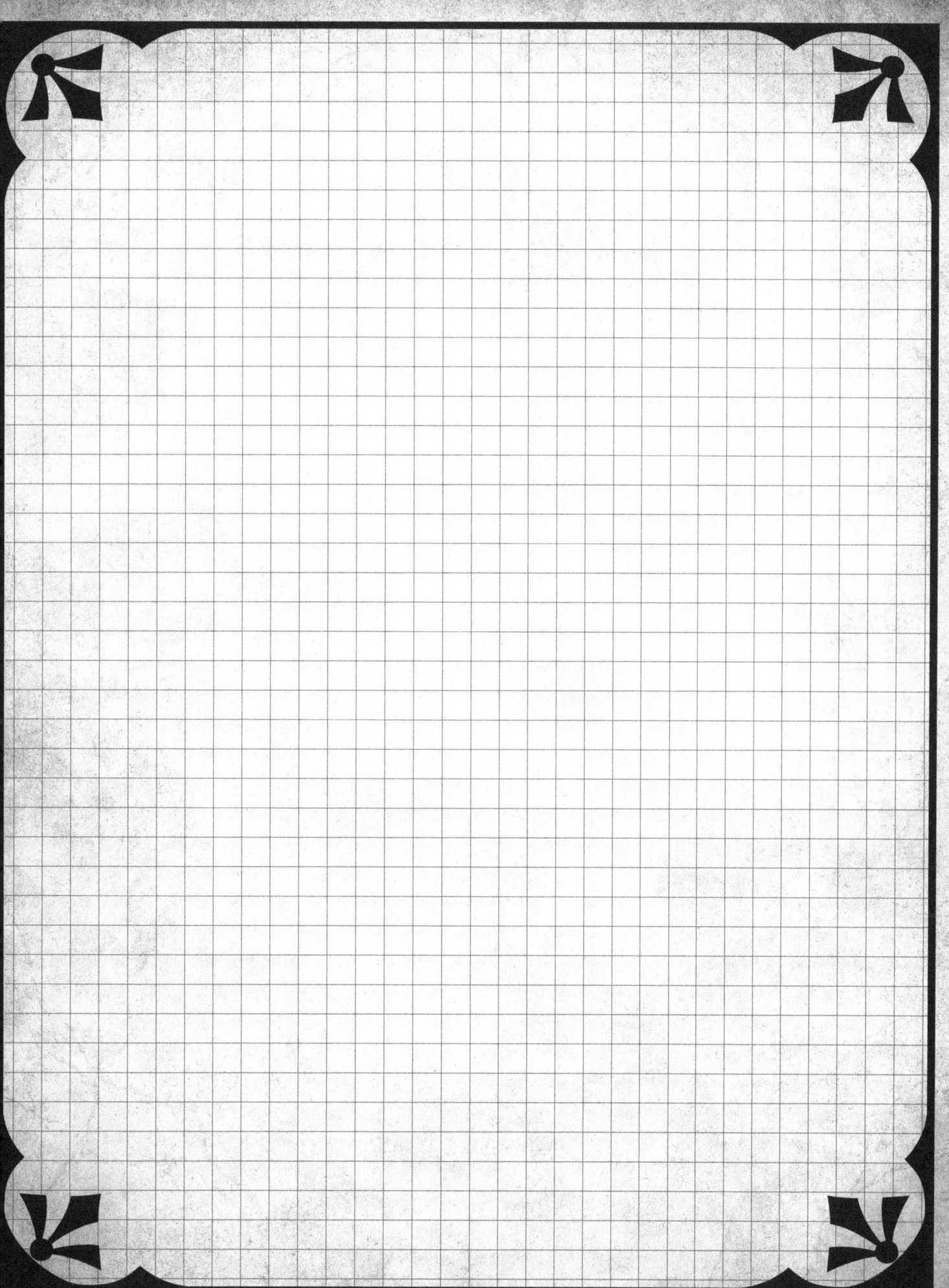

Notes & House Rules

Galactic Conquest Variations

MATH CONCEPTS: multiplication math facts, rectangular area.
PLAYERS: two to four.
EQUIPMENT: graph paper, two six-sided or gaming dice or one deck of playing cards (face cards removed), colored markers.

How Close to 100?

(A cooperative game.)

Mark a 10 × 10 square on your graph paper, and use six-sided dice. Players take turns rolling two dice and coloring in a rectangle with those dimensions anywhere on the gameboard.

Your shared goal is to fill as many squares as you can.

How close can you get to coloring the full 100 squares before someone rolls a product that won't fit?

Galactic Blobs

Players may draw any single, connected shape that covers the area representing the product of their dice or cards. Every square of the shape must share at least one side with some other part of the shape.

For example, a player could draw the L-shaped area of a chess knight's move, but not the diagonal squares of a bishop's path because those only meet at the corners.

Warp Speed

To practice the hardest multiplication facts, remove the aces, twos, threes, and tens from your deck of cards. Use graph paper with small squares, because your conquered territory will grow rapidly without the low numbers.

Deal two cards to each player. Make a discard pile beside the deck.

On your turn, draw two more cards. Choose any two cards from your hand to make a product, laying them on the table for all to see. Fill in that rectangle, and discard the used cards.

Play until the deck runs out or until two players are forced to pass in consecutive turns.

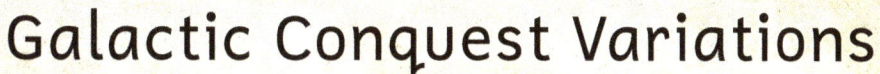

Notes & House Rules

Domino Product Train

MATH CONCEPTS: multiplication facts, numerical order, thinking ahead.
PLAYERS: two or more.
EQUIPMENT: one set of double-9, double-12, or double-15 dominoes.

Set-Up

Turn all the domino tiles face down on the table and mix them around to make the wood pile. Each player draws six tiles from the wood pile but does not look at them.

Players arrange their tiles in a row (train), as shown. When all players are ready, turn the tiles in your train face up without changing their position.

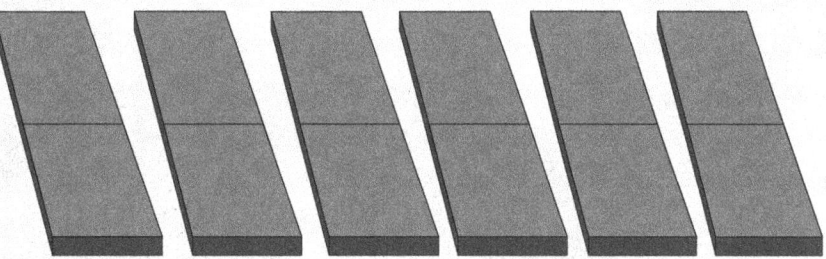

A domino number train, ready to flip and play.

How to Play

Your goal is to make the products of the two numbers on each domino tile in your train increase from left to right.

On your turn, draw one tile from the wood pile. You have two choices:

- Mix the new tile back into the pile.
- Use the new tile to replace one of yours, and then discard the old tile.

The first player to complete a train in which the products increase from left to right wins the game.

HOUSE RULE: Decide how strict you will be about the "increases from left to right" rule and repeated numbers. Do matching products (like 1 × 6 and 2 × 3) count in a valid number train? Or will the player have to keep trying for a tile to replace one of the duplicates?

Notes & House Rules

Times-Tac-Toe

MATH CONCEPTS: multiplication math facts, times tables.
PLAYERS: only two.
EQUIPMENT: blank 10 × 10 times-table chart, one deck of playing cards (face cards removed), colored markers or a set of matching tokens for each player.

Set-up

Print or draw a blank times-table chart for players to share. The chart consists of a grid with dimensions one square longer than the number of multiplication facts you wish to practice. For example, if you are using the facts from 1 × 1 up to 10 × 10, draw a square with 11 rows and columns.

Mark the multiplication symbol in the top left square of your chart, then fill the top row and left column of your blank times-table chart with the numbers 1–10, in numerical order or mixed around.

Shuffle the playing cards and place the deck face down as a draw pile where everyone can reach. If you are using colored markers, the colors must be different enough to be easily distinguished.

How to Play

On your turn, flip two cards. Multiply them and find the corresponding square on the times table. If that square is blank, write in the product with your colored marker. Or say the product aloud while you cover that square with one of your tokens. If a player writes or says the wrong answer, the other player may challenge, give the correct product, and then claim that square by coloring over your number.

Sometimes you will have a choice of two squares, but you may mark only one of them. On the other hand, if there are no remaining spaces for your product, then you lose that turn.

The first player to mark four squares touching (with no gaps) in a row horizontally, vertically, or diagonally wins the game.

Variations

For a longer game, play until someone marks five squares in a row.

If you prefer teaching the multiplication facts up to 12 × 12, you can include face cards in your deck: jack = 11, queen = 12, and king = wild card. A player who turns up a king may use any number in its place.

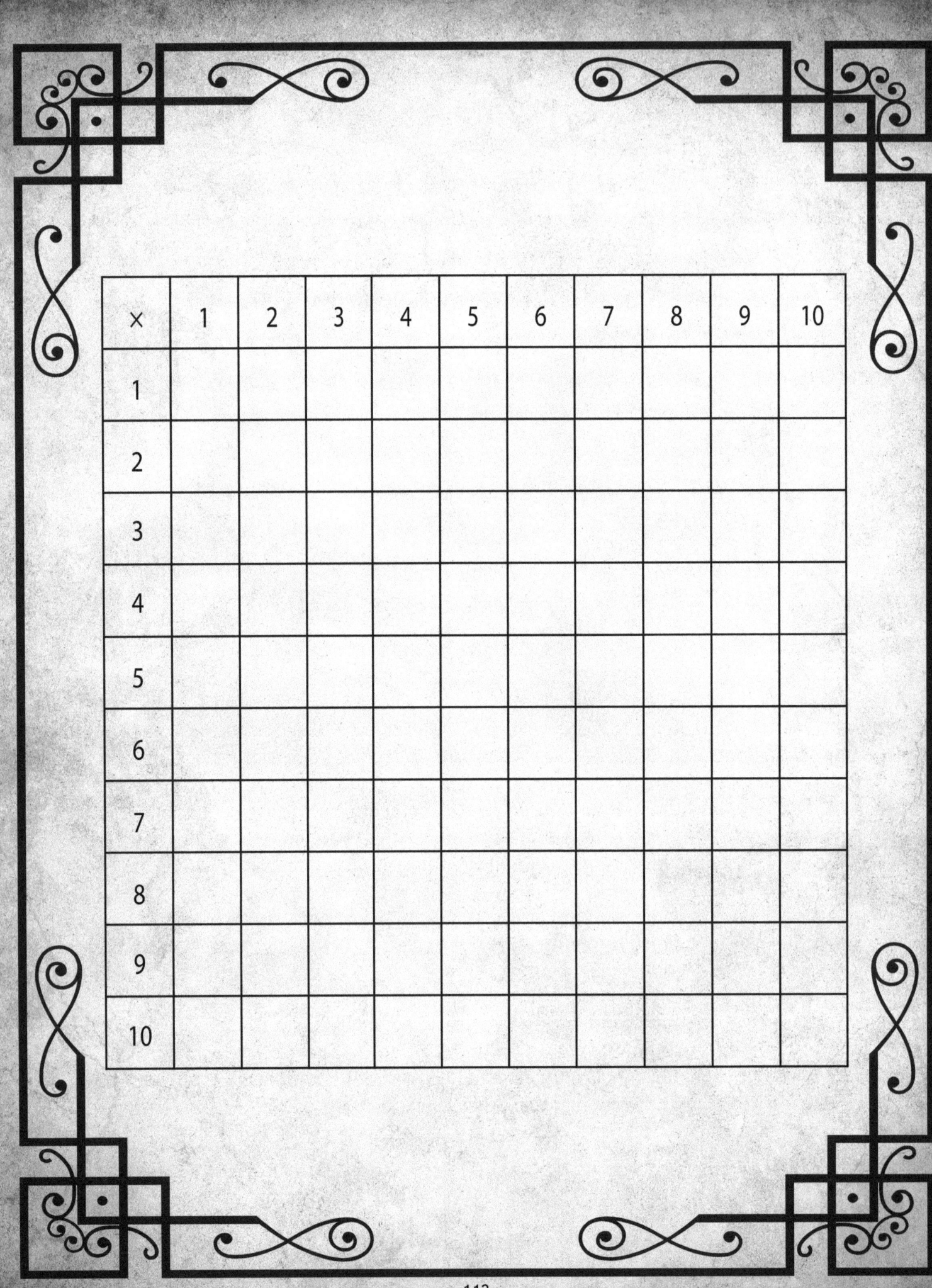

Notes & House Rules

Multiplication Gomoku

MATH CONCEPTS: multiplication math facts, times tables, strategic thinking.
PLAYERS: only two.
EQUIPMENT: blank 12 × 12 times-table chart, colored markers or a set of matching tokens for each player.

Set-up

Print or draw a blank times-table chart with thirteen rows and thirteen columns.

Mark the multiplication symbol in the top left square of your chart, then fill the top row and left column of your blank times-table chart with the numbers 1–12, in numerical order or mixed around.

If you are using colored markers, the colors must be different enough to be easily distinguished.

How to Play

On your turn, choose any unclaimed square and write in the product of its row and column numbers. Or say the product aloud while you cover that square with one of your tokens.

If a player writes or says the wrong answer, the other player may challenge, give the correct product, and then claim that square by coloring over the wrong number.

The first player to mark five squares touching (with no gaps) in a straight row—horizontally, vertically, or diagonally—wins the game.

Variation

For a tougher challenge, one player fills the top row squares with any numbers greater than five, in any order, and the other player does the same in the left-column squares.

✕	1	2	3	4	5	6	7	8	9	10	11	12
1	1	2	3	4	5	6	7	8	9	10	11	12
2	2	4	6	8	10	12	14	16	18	20	22	24
3	3	6	9	12	15	18	21	24	27	30	33	26
4	4	8	12	16	20	24	28	32	36	40	44	48
5	5	10	15	20	25	30	35	40	45	50	55	60
6	6	12	18	24	30	36	42	48	54	60	66	72
7	7	14	21	28	35	42	49	56	63	70	77	84
8	8	16	24	32	40	48	56	64	72	80	88	96
9	9	18	27	36	45	54	63	72	81	90	99	108
10	10	20	30	40	50	60	70	80	90	100	110	120
11	11	22	33	44	55	66	77	88	99	110	121	132
12	12	24	26	48	60	72	84	96	108	120	132	144

×	1	2	3	4	5	6	7	8	9	10	11	12
1												
2												
3												
4												
5												
6												
7												
8												
9												
10												
11												
12												

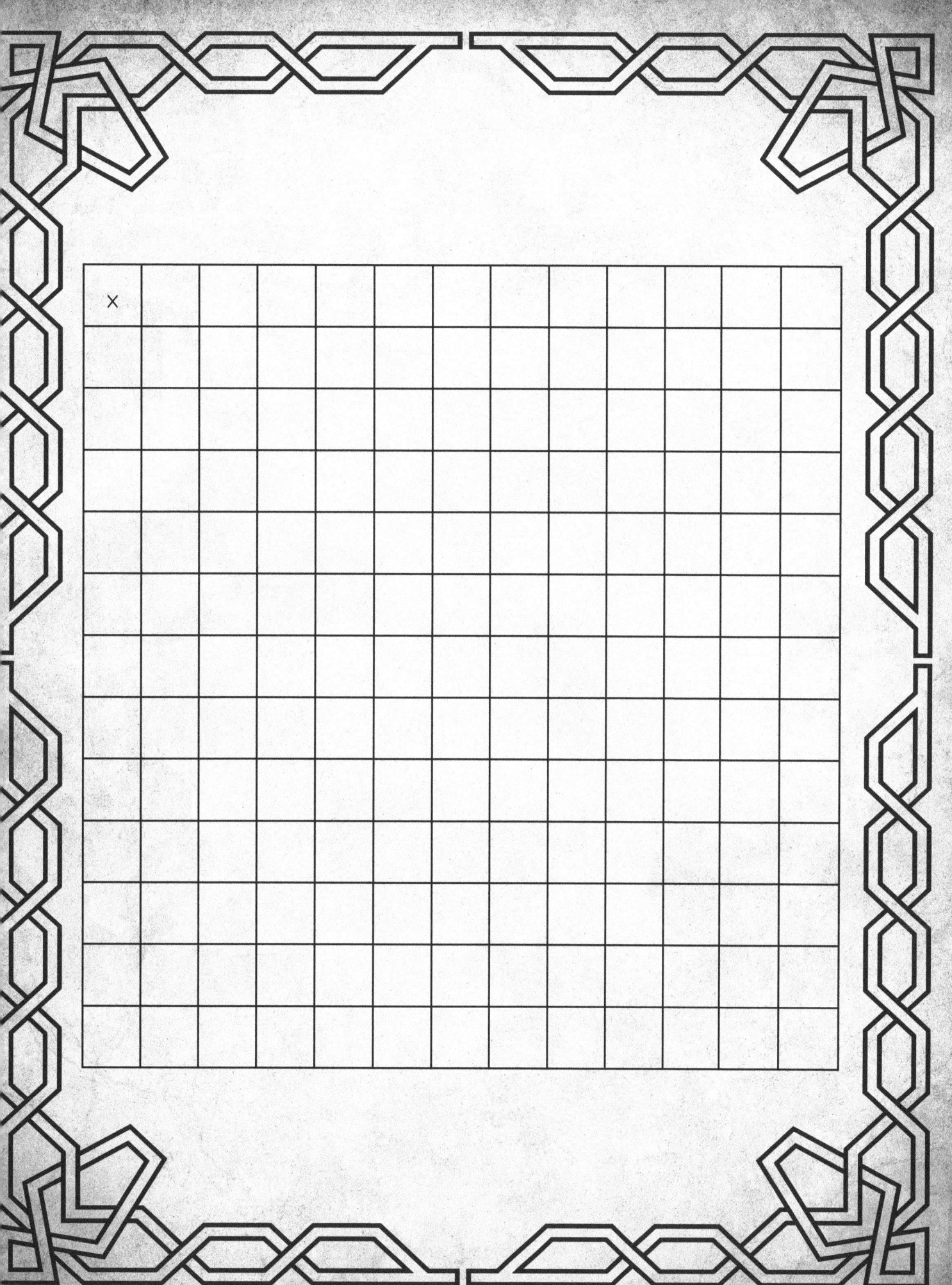

Notes & House Rules

The Product Game

MATH CONCEPTS: multiplication math facts, factors, multiples.
PLAYERS: two, or two teams.
EQUIPMENT: gameboard, colored markers or a set of matching tokens for each player, and two paperclips, glass gemstones or other small tokens for the factors.

How to Play

The first player places a paperclip on any factor at the bottom of the board. The second player places the other clip on a factor—the same or different—and then marks the product of those two numbers by coloring the square or placing a token.

On each succeeding turn, a player moves just one clip to a new number and then marks the product of those two factors. If both players agree that all possible moves have already been colored in, the player whose turn it is may make a fresh start by moving both paperclips.

Whichever player marks four (or more) squares in a row—horizontally, vertically, or diagonally—wins the game. The squares must touch each other at edges or corners, with no gaps.

If neither player can make four in a row, then the player who has the most sets of three in a row wins.

Variations

PRODUCT GAME BOXES: On your turn, you may color one line segment on any single side of the square that contains the product of your numbers. When you draw a fourth line, completing a square, then you color that square with your color and take a free turn. The game ends when either player cannot color a line to match the numbers. Whoever has colored in the most boxes wins.

PATHWAYS: One player "owns" the top and bottom of the gameboard, while the other player claims the right and left sides. The first player who can mark a pathway of squares across the board (top to bottom for one, side to side for the other) wins the game. The pathway squares must all connect by sharing a side or corner.

Build Your Own Gameboard

Players can work together to create their own gameboard:*

- On a blank sheet of paper, draw a 6 × 6 array of squares.
- In the space below the array, each player writes four factors. Factors do not have to be single-digit numbers, but may be anything the players choose to practice with.
- Then take turns writing the product of any two factors into one gameboard square. A few duplicates won't ruin the game, but try to avoid writing a product that's already been used.

When all the squares are full, you're ready to play. You may want to slip your gameboard into a clear page protector or laminate it for playing with dry-erase markers.

*For free printable gameboards, download the Multiplication & Fraction Printables at TabletopAcademy.net/free-printables.

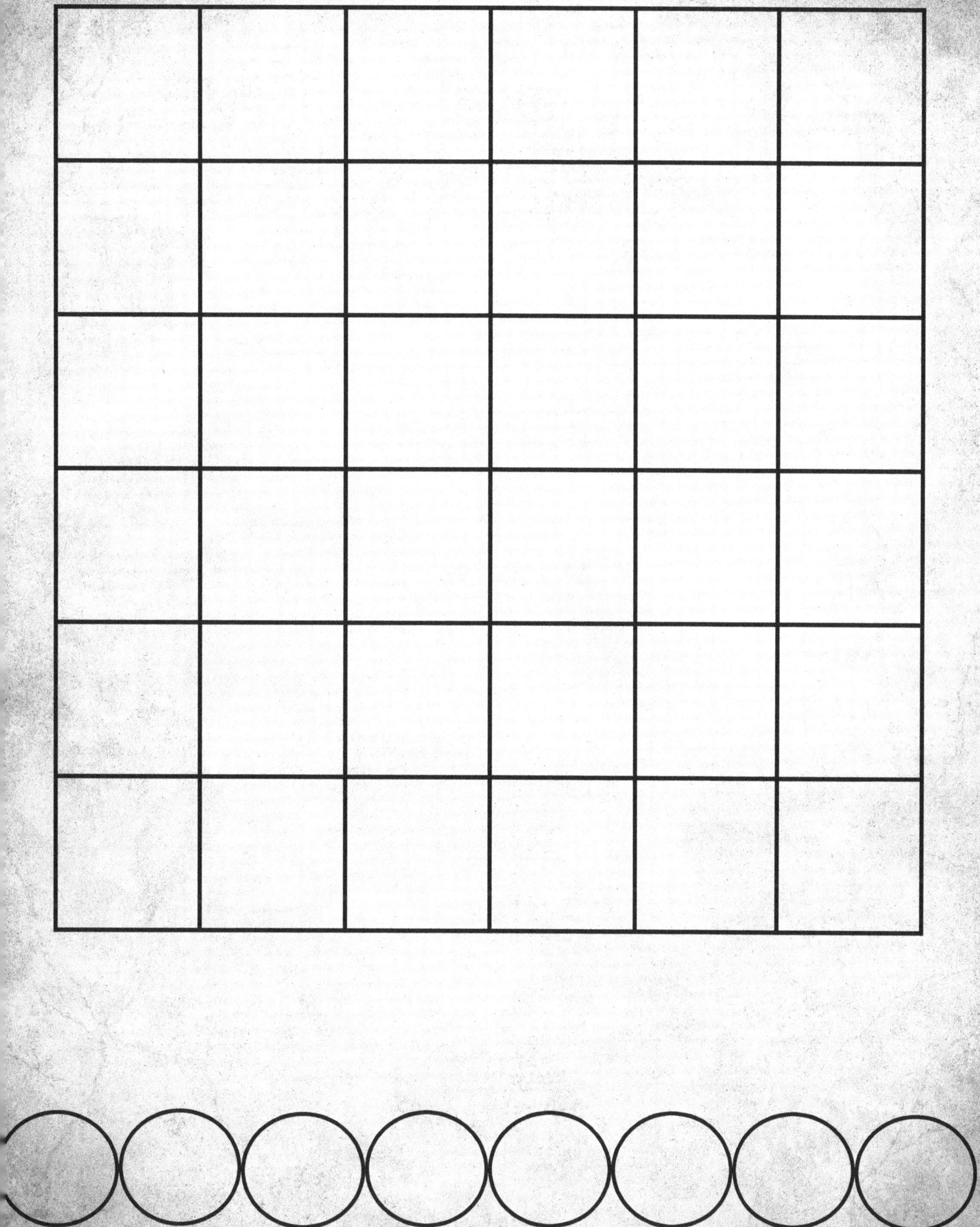

1	2	3	4	5	6
7	8	9	10	12	14
15	16	18	20	21	24
25	27	28	30	32	35
36	40	42	45	48	49
54	56	63	64	72	81

1 2 3 4 5 6 7 8 9

1	2	3	4	5	6
27	28	30	32	35	7
25	56	63	64	36	8
24	54	81	72	40	9
21	49	48	45	42	10
20	18	16	15	14	12

1 2 3 4 5 6 7 8 9

Notes & House Rules

Ultimate Multiple Tic-Tac-Toe

MATH CONCEPTS: multiplication math facts, times tables, thinking ahead.
PLAYERS: two, or two teams.
EQUIPMENT: printed or hand-drawn Multiple-Tac-Toe chart, colored markers, two paperclips, glass gemstones or other small tokens for the factors.

Set-Up

If you are starting with a blank Multiple-Tac-Toe chart, give players time to write in the multiples in pencil. Each small Tic-Tac-Toe board contains the first nine multiples of one counting number, which are the answers in that number's times table. As they fill in these numbers, players internalize the structure of the gameboard, which makes playing the game go smoothly.

How to Play

The player who is marking O goes first, placing a paperclip on any factor at the bottom of the board. The second player places the other clip on a factor—the same or different—and then marks an X on the product of those two numbers. On each succeeding turn, a player moves just one clip to a new number and then marks the product of those two factors.

Early in the game, players may choose from more than one square—for instance, for 3 × 7, the player may mark the times-three or times-seven board. As the game progresses, options will grow increasingly limited.

If the only unmarked square is on a small Tic-Tac-Toe board that has already been won, the player must still mark there. But if all possible moves have already been marked, the player whose turn it is may make a fresh start by moving both paperclips.

If you win one of the small Tic-Tac-Toe boards, mark over the whole thing with a large X or O. If a small board ends in a draw, mark it with a large C for "cat's game"—and that counts as a win for either player.

The first player to claim three large squares (that is, three of the small boards) in a row wins the game.

With a cat's game acting as a wild card, it's possible that both players can make a row in the same turn: double-win!

Variation

FAST GAME: Players mark all the products of their factor numbers at once, wherever they appear on the board. So a player who has the paperclips on two and nine will mark every 18 on the whole gameboard.

1	2	3	2	4	6	3	6	9
4	5	6	8	10	12	12	15	18
7	8	9	14	16	18	21	24	27
4	8	12	5	10	15	6	12	18
16	20	24	20	25	30	24	30	36
28	32	36	35	40	45	42	48	54
7	14	21	8	16	24	9	18	27
28	35	42	32	40	48	36	45	54
49	56	63	56	64	72	63	72	81

1 2 3 4 5 6 7 8 9

1 2 3 4 5 6 7 8 9

For More Information

Galactic Conquest

The "How Close to 100?" variation comes from Jo Boaler's YouCubed website, a resource for teaching math while promoting a growth mindset.

- ♦ youcubed.org/tasks/how-close-to-100

The Product Game

Marilyn Burns posted the Pathways game—which is a cross between The Product Game and the strategy game Hex—on her Math Blog, along with a lesson plan and several gameboards to choose from.

- ♦ marilynburnsmath.com/games/the-game-of-pathways

The Boxes game is based on Joshua Greene's post "Dots and Boxes Variation" on Three J's Learning blog. And Greene posed several questions prompting players to examine and extend The Product Game in "Times Square Variations."

- ♦ 3jlearneng.blogspot.com/2016/02/dots-and-boxes-variation.html
- ♦ 3jlearneng.blogspot.com/2015/11/times-square-variations-math-games.html

Ultimate Multiple-Tac-Toe

Ben Orlin wrote about Ultimate Tic-Tac-Toe on his delightful Math with Bad Drawings blog. Federico Chialvo modified The Product Game to create the Fast Game variation, and then wrote a follow-up post analyzing how common factors and multiples affect the strategy of his game. Then Orlin wrapped up the discussion with a post about the difference between a puzzle and a game.

- ♦ mathwithbaddrawings.com/2013/06/16/ultimate-tic-tac-toe
- ♦ artofmathstudio.wordpress.com/2013/08/30/multiplication-tic-tac-toe
- ♦ artofmathstudio.wordpress.com/2013/10/31/revisiting-multiplication-tic-tac-toe-common-factors-and-multiples
- ♦ mathwithbaddrawings.com/2013/11/18/tic-tac-toe-puzzles-and-the-difference-between-a-puzzle-and-a-game

The Great Escape

A World War II Adventure Math Game

Notes & House Rules

The Great Escape

Math Concepts: multiplication, division, factors and multiples, odd and even, prime numbers, square numbers, cubic numbers.
Players: two to four.
Equipment: hundred chart, playing cards, different colored pencils or markers.

Set-Up

You are a flight lieutenant in the Royal Air Force, captured by the Axis powers during World War II and held in a prisoner-of-war camp. Your mission is to dig an escape tunnel from your barracks to the neighboring woods.

Players share a hundred chart. Each player needs a distinctly colored pencil or marker. Spread the playing cards face down on the table as a fishing pond.

How to Play

Players take turns drawing a card from the fishing pond and coloring a matching number square on the hundred chart, according to the Secret Number Code on the next page.

Each player's first square must be on a vertical or horizontal outer edge of the gameboard. Circle that starting position with your color. After that, you may color in any unclaimed square.

If there is no number square that matches your card, you lose that turn. For instance, there are only four cubic numbers on the hundred chart. After those are all claimed, kings become "lose a turn" cards.

At the end of your turn, mix your card back into the fishing pond.

Your goal is to complete a connected path of squares (meeting at sides or corners) from your starting position to the opposite side of the board. The first player to finish such a path escapes from the prison camp and wins the game.

Variations

With four players, pair up with the player coming from the side of the gameboard opposite yours to create your joint tunnel.

For a more difficult challenge, players must tunnel directly from one square to the next. Each number you mark must touch one of your previously colored squares at a side or corner.

Or for a simpler game with no blocking, give each player a separate hundred chart gameboard. In this variation, tunnel squares must be attached at their sides; squares meeting only at a corner do not count as connected. To speed up the game, all players may draw cards and choose squares at the same time.

Secret Number Code

Color a number square that matches your card, according to the following code:

Ace = any odd number

2 = any even number

3–10 = multiple of the card number

Jack = prime number

Queen = square number

King = cubic number

Joker = multiple of one

(All whole numbers are a multiple of one, so jokers are wild cards.)

1	2	3	4	5	6	7	8	9	10
11	12	13	14	15	16	17	18	19	20
21	22	23	24	25	26	27	28	29	30
31	32	33	34	35	36	37	38	39	40
41	42	43	44	45	46	47	48	49	50
51	52	53	54	55	56	57	58	59	60
61	62	63	64	65	66	67	68	69	70
71	72	73	74	75	76	77	78	79	80
81	82	83	84	85	86	87	88	89	90
91	92	93	94	95	96	97	98	99	100

Notes & House Rules

Team Great Escape

Math Concepts: multiplication, division, factors and multiples, odd and even, prime numbers, square numbers, cubic numbers.
Players: two to four (a cooperative game).
Equipment: printed hundred chart, playing cards (with jokers), pencils or markers.

Set-Up

You are a flight lieutenant in the Royal Air Force, captured by the Axis powers during World War II and held in a prisoner-of-war camp. You're working with your fellow prisoners to escape before the Nazi guards discover your tunnel.

Print a single hundred chart for players to share. Spread the playing cards face down on the table as a fishing pond.

Place the Guard Towers

Each player draws two number cards. Tens count as zero for this part of the game. If you get a face card or joker, draw again.

Arrange your two cards to make a two-digit number. For example, if the cards are 3 and 8, you could make 38 or 83. Place a guard tower on that square by drawing an X with a circle around it.

Mix all the cards back into the fishing pond.

How to Play

Then players take turns drawing two cards from the fishing pond and marking two number squares—one for the escape tunnel and one for a patrolling Nazi guard—according to the Secret Number Code:

- Ace = any odd number
- 2 = any even number
- 3–10 = multiple of the card number
- Jack = prime number
- Queen = square number
- King = cubic number
- Joker = multiple of one

(All whole numbers are a multiple of one, so jokers are wild cards.)

Digging the Tunnel

Choose which of your two cards to use for the escape tunnel and color in a square that matches it. Then find a square that matches the other card and mark it with an X to represent the guard. Once a square is marked, it cannot be used in future turns.

Each player's initial escape tunnel square must be on a vertical or horizontal outer edge of the gameboard. The first player may choose any edge. The second player must start on the opposite side of the board from the first. The third chooses either unmarked edge, and the fourth player takes what's left.

If you can't find a square on your edge that matches either of your cards, you must still mark a Nazi guard position.

Subsequent Moves

After your initial move, you may build a tunnel in any unclaimed square. If one of the players hasn't made it onto the board, you may mark a square on their edge to let them in.

If one of your cards has no matching number square, use it for the guard (sleeping on patrol) and the valid card for your tunnel. But if both cards have no match, you lose that turn. For instance, there are only four cubic numbers on the hundred chart. After those are all claimed, kings become "lose a turn" cards.

Endgame

Your goal is to dig a tunnel of squares that lets all players escape. Squares must be attached at their sides; squares meeting only at a corner do not count as connected. If the Nazi patrol blocks your path, you lose the game. But if you complete a tunnel that connects all players' initial squares, you win.

Towers give the Nazi guards a head start, but crafty prisoners can work around them.

Which card would you use for your tunnel, and where would you place a guard?

Secret Number Code

Color a number square that matches your card, according to the following code:

Ace = any odd number

2 = any even number

3–10 = multiple of the card number

Jack = prime number

Queen = square number

King = cubic number

Joker = multiple of one

(All whole numbers are a multiple of one, so jokers are wild cards.)

1	2	3	4	5	6	7	8	9	10
11	12	13	14	15	16	17	18	19	20
21	22	23	24	25	26	27	28	29	30
31	32	33	34	35	36	37	38	39	40
41	42	43	44	45	46	47	48	49	50
51	52	53	54	55	56	57	58	59	60
61	62	63	64	65	66	67	68	69	70
71	72	73	74	75	76	77	78	79	80
81	82	83	84	85	86	87	88	89	90
91	92	93	94	95	96	97	98	99	100

For More Information

Australian journalist and fighter pilot Paul Brickhill wrote the true story of perseverance, heroism, and tragedy during World War II: *The Great Escape*, W. W. Norton & Co, 1950.

This game comes from Marina Singh's MathCurious blog:

- mathcurious.com/2020/04/02/the-great-escape

In Singh's original The Great Escape game, players compete against each other. But in real life, more than six hundred prisoners worked together to construct three tunnels under Stalag Luft III. One tunnel made it through.

Seventy-six men escaped before the guards discovered the tunnel, filled it in, and launched a massive manhunt. Unfortunately, only three of the escapees made it all the way across Germany to safety.

But the project was important, even for those left behind. RAF pilot Jack Lyon explains:

"It did a lot for morale, particularly for those prisoners who'd been there for a long time. They felt they were able to contribute something, even if they weren't able to get out. They felt they could help in some way and trust me, in prison camps, morale is very important."

—Jack Lyon

From a speech at a Royal Air Force Benevolent Fund event to mark the 70th anniversary of the mission. Quoted by Lucy Crossley:

- dailymail.co.uk/news/article-2579916/Failure-The-Great-Escape-worth-human-cost-boosted-morale-Nazi-prison-camps-say-survivors.html

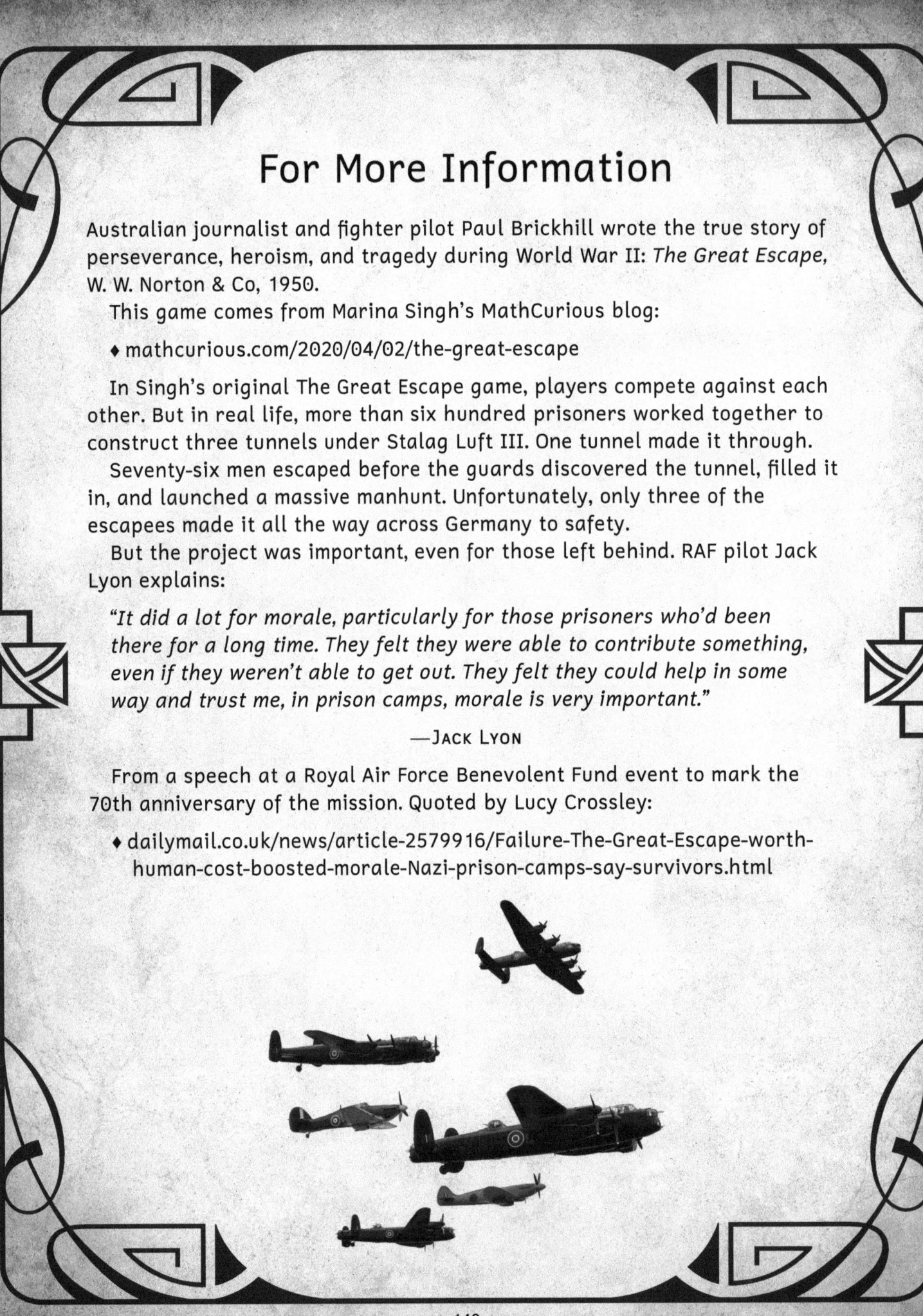

TABLETOP MATH GAMES COLLECTION

INTEGER GAMES

6 WAYS TO PLAY MATH WITH OLDER STUDENTS

Notes & House Rules

Consecutive Capture

MATH CONCEPTS: integers, number line.
PLAYERS: two or more.
EQUIPMENT: playing cards, clothesline rope and clothespins (optional).

Set-Up

You need a wide horizontal area on your table to arrange your cards into a number line. Or stretch a length of rope between two chairs or two hooks on a wall, and provide a basket full of clothespins.

Agree on which color represents negative numbers. At our house, we play with accounting colors: Black cards are positive numbers, and red cards count as negative. But some families like to use the electrician's standard that red is positive, black negative.

In this game, an ace may be used as one or twelve, player's choice. The number cards represent their face value. Count the jack as eleven, queen as zero, and king as a wild card that can be any number you wish.

How to Play

Deal three cards to each player. Set the rest of the deck on the table as a draw pile.

On your turn, place one of your cards in its proper position on the number line. If two cards have the same value, stack or clip them together.

If your card makes a series of three or more consecutive integers, you capture all the consecutive cards. Remove them from the clothesline. But if there are two cards with the same value, only take one of those. Add the captured cards to your score pile.

Finally, draw a new card to replenish your hand.

When the deck is gone, turns continue until all cards are played. Whoever collects the most cards wins.

Notes & House Rules

Strike It Out

MATH CONCEPTS: integer addition and subtraction.
PLAYERS: only two.
EQUIPMENT: blank paper, pencils or markers.

How to Play

Draw a number line and label the numbers from −10 to +10. The first player strikes two numbers (draws a line through them) and circles their sum, saying the equation.

Then the second player must strike out the number the first player circled and one unused number, and then circle either their sum or their difference—which also must be an unused number. Be sure to say your equation out loud, so the other player can check your mental math.

Once a number is struck out by either player, it can't be used again.

The first player must add, but after the first turn you may add or subtract. Continue to take turns striking two numbers (one of which must be the answer from the previous turn) and circling their sum or difference.

Eventually the numbers run out or the remaining numbers are impossible to reach. Whoever makes the last legal move wins the game.

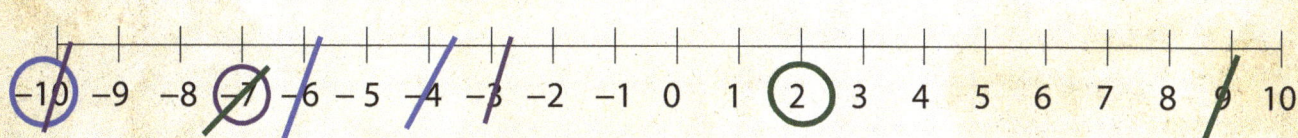

$$-4 + -6 = -10$$
$$-10 - (-3) = -7$$
$$-7 + 9 = 2$$

A sample game of Strike It Out. I've written equations to show each turn's play, but players may explain their moves orally.

Notes & House Rules

Integer Solitaire

MATH CONCEPTS: integer addition and subtraction.
PLAYERS: one or more (a cooperative game).
EQUIPMENT: playing cards, large sheet of poster board (optional).

Set-Up

Draw the four equations (two with addition, and two with subtraction) on a sheet of poster board, large enough that playing cards fit in the blanks.

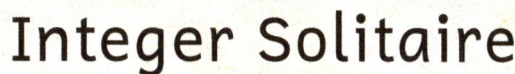

If you don't have poster board, draw the equations on plain paper so players can write in their numbers. You may want to slip this gameboard into a clear page protector or laminate the paper for use with dry-erase markers, to make it easier to move the numbers around.

Agree on which color represents negative numbers. Aces count as one, number cards at their face value. The jack, queen, and king are eleven, twelve, and thirteen, respectively.

How to Play

Turn up eighteen cards and set the rest of the deck aside. Arrange these cards in the boxes on your gameboard. There are only fourteen blanks, so you won't use all the cards.

Can you make four true equations? If so, you win.

Variation

If you succeed with eighteen cards, try the game again with seventeen. Can you do it with sixteen cards? Is fifteen enough to make it work?

The four equations of Integer Solitaire, shown with a winning arrangement of cards.

Notes & House Rules

Grid Fight

MATH CONCEPTS: integer multiplication, area model for multiplication.
PLAYERS: two players or two teams.
EQUIPMENT: square grid paper, playing cards, pencils or colored markers.

Set-Up

Players share a gameboard on square grid paper. Outline two 12 × 12 squares separated by a "zero" line.

Label one square positive (+) and the other negative (−). Choose which player or team will use the positive side of the gameboard grid and who takes the negatives.

Remove the face cards from your deck. Agree on which color represents negative numbers.

How to Play

Deal three cards to each player. Place the rest of the deck face down as a draw pile.

On each turn, both players make their move at the same time. Choose one card to play and hold it face down on the table in front of you. When both players are ready, turn up the chosen cards and multiply the numbers.

If the product is greater than zero, the positive player colors a rectangle of that size (with the two numbers as length and width) on the positive side of the grid. If the product is negative, that player colors a rectangle on the negative side.

Your goal is to fill in as many complete rows as you can, as in the game Tetris. But if you can't fit your rectangle on the grid, you lose that turn.

Both players discard their used cards and draw to replenish their hands.

Play twelve turns. Whoever fills in the most rows wins the game.

Variation

Do you want a longer game? Play on two gameboards at once. Players may color their rectangle in either square (on their own side of the zero line) wherever they can fit it in. The game continues until the grids are so full that players lose three turns in a row.

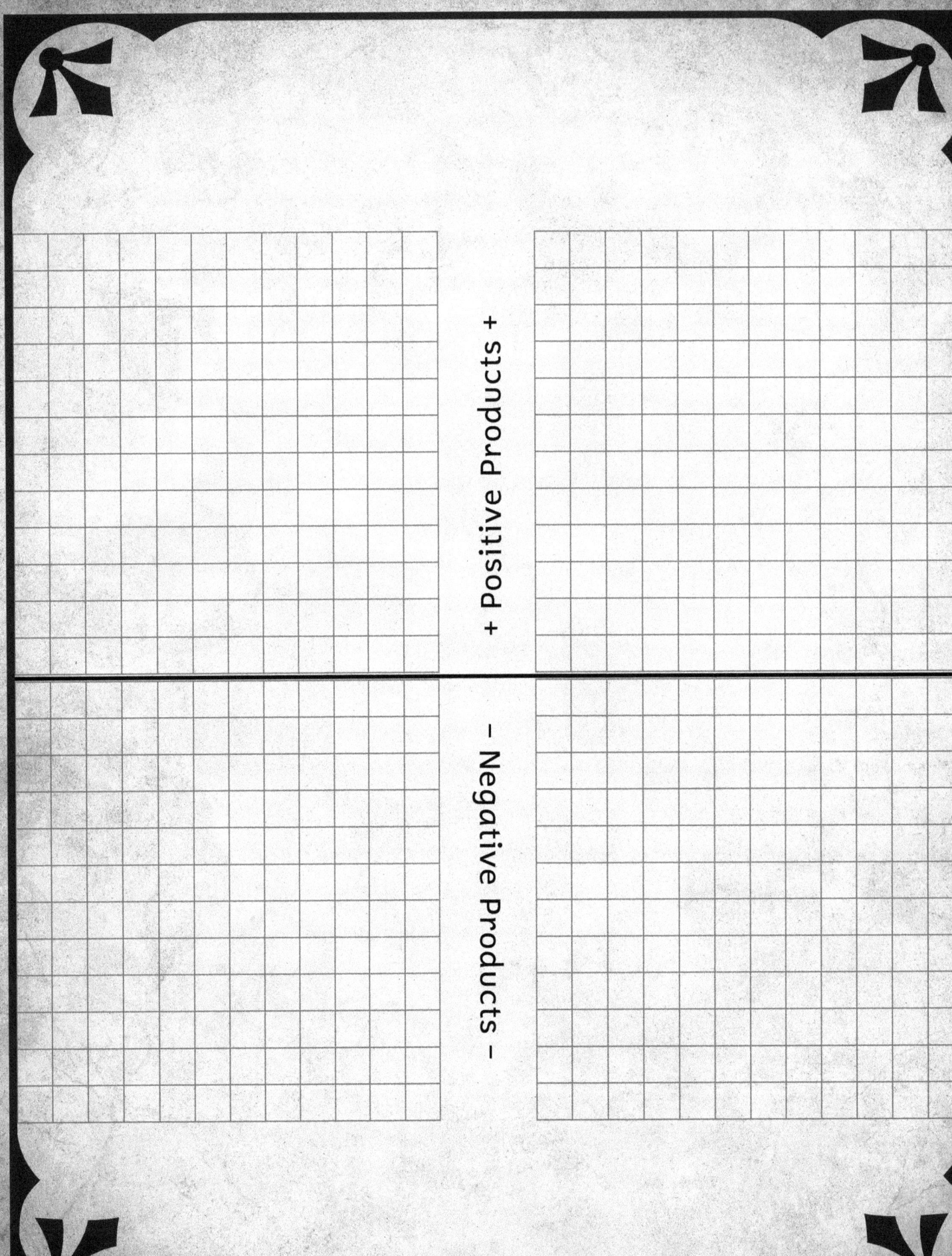

+ Positive Products +

− Negative Products −

Notes & House Rules

For free printable gameboards, download the Prealgebra & Geometry Printables at TabletopAcademy.net/free-printables.

Honeycomb

MATH CONCEPTS: integer addition, integer multiplication.
PLAYERS: two players or two teams.
EQUIPMENT: hexagon grid paper, deck of playing cards, pencils or markers.

Set-Up

Outline a gameboard on hexagon grid paper for the players to share: one large hexagon with three smaller hex spaces on each side. Or use a full-page hexagon grid for a longer game.

Agree which color card represents negative numbers. Remove the face cards and all numbers greater than six. Spread the remaining cards face down to make a fishing pond.

How to Play

Decide which player will take negative numbers, and the other takes positive. The negative player goes first.

On your turn, draw a card from the fishing pond. With the resulting positive or negative number, you can either:

- Add that value by writing it into any open hexagon.
- Multiply an existing hexagon number by that value to create a straight line of adjacent hexagons.

You multiply by *replicating* the existing number, copying it into open spaces connected to the original. The line must include as many hexagons as the value you are multiplying by. And if your number is negative, change the sign of the numbers in the row. You can multiply only if there are enough open spaces to make a straight line. Multiplication rows may not turn.

Suppose you drew −4. You might...

- Add a −4 in any unclaimed hexagon. Or...
- Notice that one hexagon has a 5 with plenty of blank spaces around it. Write three more 5s in a row and put minus signs on them all, for a total of four −5s. Or...
- Notice a space with −1 and multiply it into a row of four +1s.

After your turn, place your card back in the fishing pond and mix thoroughly.

Honeycomb Endgame

Play until the gameboard is filled. If the sum of all the numbers on the gameboard is less than zero, the negative player wins. If the sum is greater than zero, the positive player wins.

The losing player gets to choose positive or negative for the next game.

Variations

If you're playing on a full-page hexagon grid, you may wish to include all the number cards in your fishing pond. Remove the face cards, or count them as wild cards.

HOUSE RULE: Do you wish multiplication lines could turn to fit on the gameboard? Let multiplication run like a curvy snake, as long as all the replicated copies connect to the original number with no gaps.

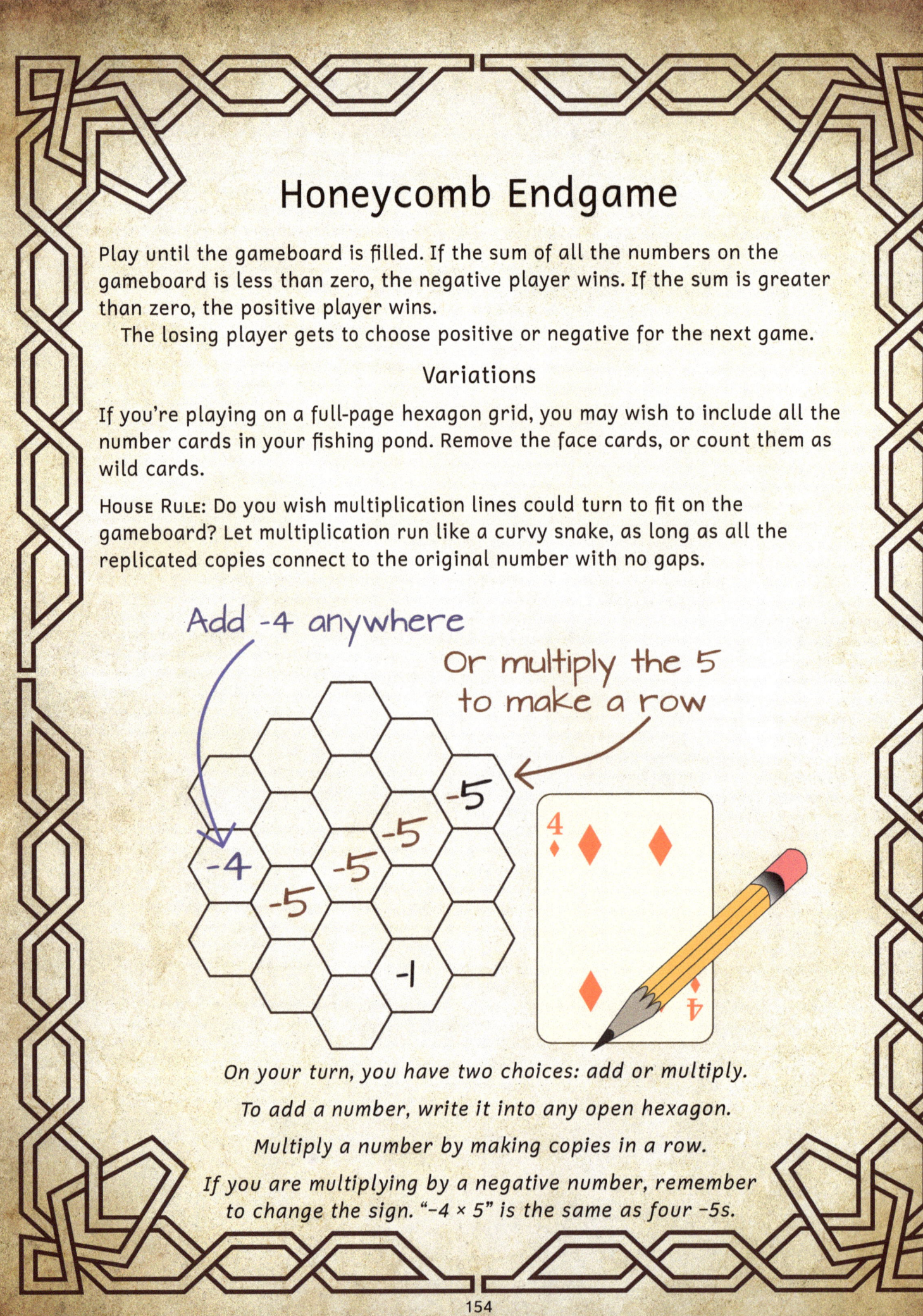

On your turn, you have two choices: add or multiply.

To add a number, write it into any open hexagon.

Multiply a number by making copies in a row.

If you are multiplying by a negative number, remember to change the sign. "−4 × 5" is the same as four −5s.

Notes & House Rules

Four in a Row Integers

MATH CONCEPTS: integer multiplication, factors, multiples, algebraic multiplication.
PLAYERS: two players or two teams.
EQUIPMENT: gameboard, colored markers, two paperclips, glass gemstones or other small tokens for the factors.

Set-Up

Choose a Four-in-a-Row gameboard for the players to share. Or have players work together to create their own gameboard:

- ♦ On a blank sheet of paper, draw a 6 × 6 array of squares.

- ♦ In the space below the array, each player writes four factors. Factors may be positive or negative and may include fractions, decimals, or even algebraic variables.

- ♦ Then take turns writing the product of any two factors into one gameboard square. A few duplicates won't ruin the game, but try to avoid writing a product that's already been used.

When all the squares are full, you're ready to play. You may want to slip your gameboard into a clear page protector or laminate it for playing with dry-erase markers.

How to Play

The first player places a paperclip on any factor at the bottom of the board. The second player places the other clip on a factor—the same or different—and then marks the product of those two numbers by coloring the square.

On each succeeding turn, a player shifts one paperclip to a new number and then marks the product of those two factors. If both players agree there are no possible moves, the player whose turn it is makes a fresh start by changing both clips.

Whichever player marks four (or more) squares in a straight row—horizontally, vertically, or diagonally—wins the game. The squares must touch each other at edges or corners, with no gaps.

If neither player connects four, then the player who has the most sets of three in a row wins.

-36	-30	-25	-24	-20	-18
-16	-15	-12	-10	-9	-8
-6	-5	-4	-3	-2	-1
1	2	3	4	5	6
8	9	10	12	15	16
18	20	24	25	30	36

9	12	15	16	18	20
-18	-16	-15	-12	-9	24
-20	24	-25	-30	-36	25
-24	36	30	25	-24	30
-25	-15	-16	-18	-20	36
-30	-36	20	18	16	15

(-6) (-5) (-4) (-3) (3) (4) (5) (6)

25	30	-35	-40	45	50
-55	-60	36	-42	-48	54
60	-66	-72	49	56	-63
-70	77	84	64	-72	-80
88	96	81	90	-99	-108
100	-110	-120	121	132	144

For More Information

Strike It Out

Strike It Out is from the Nrich Maths website, which is full of mathematical activities and resources for teachers and school-age students. At age seven, John Domoradzki (son of author Amber Domoradzki) suggested extending the game into negative numbers.

Integer Solitaire

Math instructor Kent Haines, who created this puzzle game, writes:
  "I love this game because it doesn't require a lot of supplies, can be played in fifteen minutes, and remains challenging even after a student has mastered integer addition and subtraction.
  "I picked the starting amount of cards on intuition. I have played this game for five years with dozens of students, and I have yet to see a combination of eighteen cards that is unsolvable."

- ♦ kenthaines.com/blog/2016/2/19/integer-solitaire

Gridfight, Honeycomb

I owe these games (and many others!) to the amazing John Golden. Visit his Math Hombre Games webpage for more ideas:

- ♦ mathhombre.blogspot.com/p/games.html
- ♦ mathhombre.blogspot.com/2011/02/integer-games.html

Four in a Row Integers

I first saw this type of four-in-a-row game in 1987 on the old *Square One TV* show. The basic integers gameboard comes from John Golden's "Integer Games" blog post.
  Bill Lombard and Brad Fulton published the challenge gameboard with larger absolute values in *Simply Great Math Activities: Algebra Readiness* from Teacher to Teacher Press, along with this math journal prompt:

  "A team has asked for your advice. Their opponent can win by covering the positive 36. Tell them which moves would be good ones and which moves they should avoid. Explain your reasoning to them."

TABLETOP MATH GAMES COLLECTION

MENTAL MATH MASTERY

7 WAYS TO PLAY MATH WITH OLDER STUDENTS

Notes & House Rules

Operations

MATH CONCEPTS: addition, subtraction, multiplication, division, order of operations, integers, fractions.
PLAYERS: two or more.
EQUIPMENT: regular playing cards, blank index cards or card-sized pieces of paper, pencils or markers. Calculator optional.

Set-Up

Give each player a stack of ten index cards or blank squares of paper. On each card, write the symbol for an arithmetic operation (+, −, ×, ÷) and an integer. For example, you may create cards that say "+7" or "÷12" or "×(−2)." Make all of your own cards different, but it doesn't matter if more than one player accidentally creates the same card.

Combine all the players' operation cards together in one deck. Save these cards for future games, adding a few more cards each time you play.

Agree on which color of playing card represents negative numbers. Aces count as one, face cards as twelve. (To make division easier, we skip eleven.)

How to Play

Keep the two decks of cards separate. Set the playing cards face down on the table. For the initial play, deal five operation cards to each player.

Then the dealer calls "high" or "low" for which value wins that hand. Finally, turn up the top card from the playing card deck to determine that hand's starting number N.

Players choose three operation cards from their hands and lay them down in sequence to form a mathematical expression beginning with the number N. Players each calculate the value of their own expression, following the standard order of operations.

Scoring

The expression with the greatest (or least) value scores one point. If more than one player has the same winning value, they each get a point.

Other players may want to examine the winner's calculation to make sure they agree, because order of operations can be confusing even for adults.

All players discard their used operations, keeping their two unplayed cards for the next hand. Pass the operations deck to a new dealer, who gives each player three cards to replenish their hands. If there aren't enough operation cards left, shuffle the discards back into the deck before dealing.

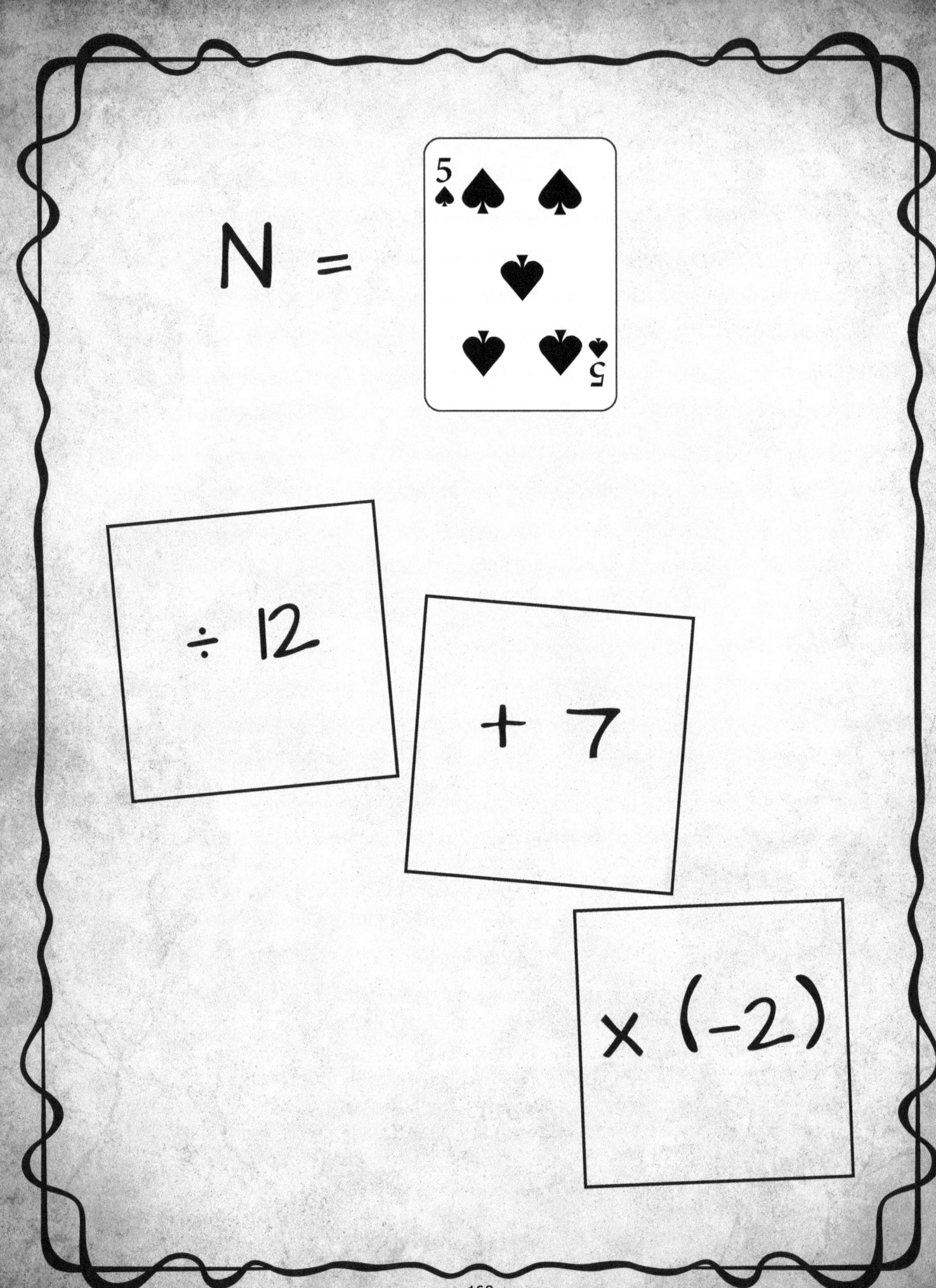

Operations Example

If the starting number is N = 5, a player might arrange the cards +7, ÷12, and ×(-2) to create expressions like:

$$5 + 7 \div 12 \times (-2)$$
$$= 5 + {}^{7}/_{12} \times (-2)$$
$$= 5 + (-{}^{14}/_{12})$$
$$= 3\, {}^{5}/_{6}$$

or

$$5 \times (-2) + 7 \div 12$$
$$= -10 + {}^{7}/_{12}$$
$$= -9\, {}^{5}/_{12}$$

or

$$5 \div 12 + 7 \times (-2)$$
$$= {}^{5}/_{12} - 14$$
$$= -13\, {}^{7}/_{12}$$

etc.

Endgame

Turns continue until everyone has a chance to deal or until players agree to stop. Whoever has the highest score wins the game. In case of tied scores, players may deal one more hand as a final showdown.

Variations

EXPONENTS: If you wish, you may allow exponent cards in your operations deck. Use the caret symbol "^" (which means "to the power of") and an integer.

HOUSE RULE: Does your family want grouping symbols to add variety to their expressions? Create a set of parenthesis and bracket cards so a player can turn an expression like:

$$5 + 7 \div 12 \times (-2)$$

... into a friendlier calculation such as:

$$[5 + 7] \div 12 \times (-2)$$

Set these handy cards on the table where all players may reach them as needed.

Notes & House Rules

Exponent Pickle

MATH CONCEPTS: integers, powers (exponents), orders of magnitude.
PLAYERS: any number.
EQUIPMENT: deck of playing cards, whiteboard and markers for each player, or pencils and blank paper. Calculator recommended.

Set-Up

Each player draws a path with ten spaces big enough to write in. Exponent Pickle paths may be utilitarian or creative: a simple row of boxes, a curvy chain of circles, a series of stair steps, a caterpillar of ovals with legs, or a string of flowers with open centers. But every path needs to have ten spaces with a clear beginning and end.

If you draw the paths on paper, you can laminate these drawings or slip them into sheet protectors for repeated play. But if you make a new drawing each time, then the game can express the players' personalities as their artistic skills develop.

Agree on which color of cards represent negative numbers. Aces are worth one, and face cards are zeros. Spread the cards out face down to make a fishing pond.

How to Play

Players may take turns, or everyone may draw cards at the same time. If the whole group plays at once, all players must wait until the others discard before drawing for the next turn.

Draw two cards from the fishing pond. If the cards are the same color, choose a third card. If they're still all the same, keep drawing until you get a card of the opposite color.

Now, arrange your cards to make a math expression with an exponent. If you only have two cards, use one number as the base and the other as exponent. With more than two cards, get creative. Optional but very helpful: Use a calculator to test your different possibilities.

Say your expression and write it into one space on your path. (Also write the value of your expression beside it, if you needed the calculator.) Always make sure the values increase from the beginning of your path to the end. If you can't create an expression that keeps the least-to-greatest pattern, then you miss that turn.

To end your turn, place all your cards face down into the fishing pond. Mix them in thoroughly, so the next player can't guess which card is where.

The first person to fill a path, with all the values in order from least to greatest, wins the game.

House Rules and Variations

House Rule 1

Decide how strict you will be about the "increasing order" rule and repeated values. Can a player use both $(-6)^0$ and $1^{(-5)}$ as part of a valid path? Or must the player keep trying for new cards to replace one of the equivalents?

House Rule 2

Does a poor choice early in the game cause frustration? Allow players to erase an expression from their paths, in place of taking a regular turn.

For Beginners

Play a simpler game. When players draw three cards, they must write an expression of this form:

$$A \times B^C$$

Lucky players who draw more cards may choose any three of the numbers to make an $A \times B^C$ expression.

Pickle Solitaire

Exponent Pickle is also fun as a single-player game. Can you fill your path without missing a turn?

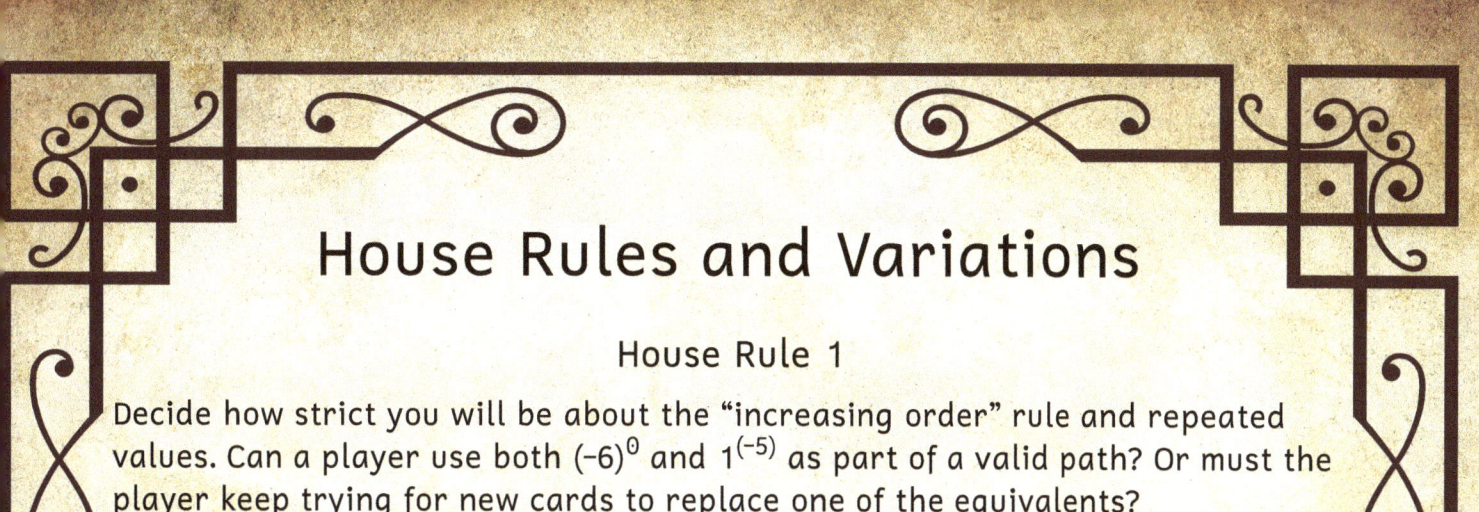

Notes & House Rules

Krypto

MATH CONCEPTS: addition, subtraction, multiplication, division, order of operations, fractions, decimals, integers, absolute value, powers and roots, factorials.
PLAYERS: any number.
EQUIPMENT: one deck of cards, pencil and paper for each player and for keeping score. Calculator optional.

Set-Up

Agree on which color represents negative numbers. Agree on the value of face cards—for example, jack = eleven, queen = twelve, king = thirteen. Or maybe you want all face cards to equal ten, while the ace can be worth the player's choice of either one or eleven.

Decide whether to set limits on the math operations allowed in your game. Or let dealers set their own limits at the beginning of each hand, before turning up any cards.

Also, agree on how long you want to play. A traditional game of Krypto runs for ten hands, but you may prefer a shorter game.

How to Play

The dealer states the allowed operations for this round and then turns up one or two cards (dealer's choice), arranging them to make the target number. For example, if the cards are 2 and 7, the target could be 27 or 72.

Leaving a space to separate them from the target number, the dealer turns up five additional cards. Players race to use these numbers in any order to create a mathematical expression equal to the target number.

You must use all five cards in your expression, and each card may be used only once by each player.

If all players agree the hand is impossible, the dealer turns up three additional cards. Players may choose any five of the eight cards (the original five plus these three) to make the target number.

Target = [2♠] [7♦]

How might you calculate 27 with the cards below?

[10♠] [4♣] [5♦] [4♥] [7♣]

One possible answer: 27 = 4 × 7 − (10 + −4 + −5).

Krypto Calculations

Calculations may include:

- The arithmetic operations +, −, ×, ÷
- Decimal points, so that 5 could become 0.5 (The zero before a decimal point is free, serving only to make the point visible. But you can't add zeros after the point to make a number like 0.05.)
- Concantenation (putting cards together to make a multi-digit number)
- Absolute values
- Square roots
- Exponents—take one card to the power of another
- Factorials
- Parentheses, brackets, or other grouping symbols
- The overhead-bar (*vinculum*) to mark a repeating decimal

You may combine these calculations in creative ways. For example, if one of the cards is −1, you could use a decimal point and a vinculum to make the repeating decimal equivalent to $-\frac{1}{9}$. Then use the absolute value, if a positive fraction is more useful in reaching the target number. And the square root can turn that positive fraction into $\frac{1}{3}$, if you wish.

Scoring

A player who figures out how to make the target number shouts "Krypto!" The player must demonstrate the calculation, either in writing or by moving the cards around, so the other players can check it.

If the calculation is valid, that player scores a point. If the same player calls Krypto on the next hand, he adds two more points to his score. The point value doubles in each succeeding hand until another player breaks the winning streak.

If the calculation was not valid, however, the player loses a point and must sit out the next hand. A player's score can go below zero.

After each hand, the deal passes to the next player. The highest score after ten hands wins the game.

Notes & House Rules

Power Krypto

MATH CONCEPTS: addition, subtraction, multiplication, division, order of operations, fractions, decimals, integers, absolute value, powers and roots, factorials.
PLAYERS: any number.
EQUIPMENT: one deck of cards, pencil and paper for each player and for keeping score. Calculator optional.

How to Play

Do you hate games that require speed? Deal cards as in the original Krypto game, then let players use any of the cards to make their expression, writing it on paper and keeping it hidden until everyone else is ready.

You may still use each card only once, as in the original Krypto game, but you do not have to use all five cards. Then players all reveal their expressions. If you had time to think of more than one expression, reveal your favorite.

If the other players agree your expression works, count the number of cards used and score two to that power. And if your expression is unique, you score double points.

For instance, suppose you used four of the five cards to calculate the target number, and no one else shared the same expression. You score $2^4 = 16$ doubled, for a total of 32 points.

The first player to reach 300 points wins the game. If more than one player passes 300 on the same turn, the highest score wins.

Variation: Road Trip Krypto

Each player needs a clipboard (or other hard surface to write on) with paper and pencil. Players take turns naming a positive or negative number, until there are eight numbers named. Numbers may be repeated. All players write these game numbers on their paper.

Then the driver names a target number. The driver also sets a time limit for the game—perhaps until the next gas station or rest area. Players try to make the target using one or more of the game numbers as many ways as they can. Each game number may be used only once per calculation. But if the game numbers include duplicates, you can use both in the same expression.

For every valid expression equal to the target, count the numbers used and score two to that power. And if your expression is unique, you score double points for that expression. High score wins.

Notes & House Rules

Krypto Insanity

Math Concepts: addition, subtraction, multiplication, division, order of operations, fractions, decimals, integers, absolute value, powers and roots, factorials.
Players: any number.
Equipment: one deck of cards (or two decks for a large group), pencil and paper for each player and for keeping score. Calculator optional.

Set-Up

Agree on which color represents negative numbers. Agree on the value of face cards—for example, jack = eleven, queen = twelve, king = thirteen. Or maybe you want all face cards to equal ten, while the ace can be worth the player's choice of either one or eleven.

Decide whether to set limits on the math operations allowed in your game.

Deal ten *hands*—sets of five cards each, face down on the table. Turn up one of the two remaining cards as the first target number. Set the final card under it, to use later.

Calculation Options

- The arithmetic operations +, −, ×, ÷
- Decimal points, so that 5 could become 0.5 (The zero before a decimal point is free, serving only to make the point visible. But you cannot add zeros after the point to make a number like 0.05.)
- Concantenation (putting cards together to make a multi-digit number)
- Absolute values
- Square roots
- Exponents—take one card to the power of another
- Factorials
- Parentheses, brackets, or other grouping symbols
- The overhead-bar (*vinculum*) to mark a repeating decimal

Ready to play Krypto Insanity.

Krypto Insanity Rules

For one complete game of Krypto Insanity, play three rounds:

- ♦ Round 1: Current face-up target number
- ♦ Round 2: Final card as target number
- ♦ Round 3: Target number is zero

Everyone plays at once. Pick up any of the face-down hands. Try to make the target number with those five cards using each card number exactly once. If you find a solution, call "Krypto."

If you cannot see a solution, turn that hand face down and pick up another. You may choose a different hand as many times as you like.

When you (or any player) call "Krypto," everyone else stops working to listen. Show your cards and explain the calculation. If all players agree it's a valid expression, set that hand of cards in your scoring pile. Then grab another hand and keep going. If your expression isn't valid, return that hand to the table and try a different set of cards.

Keep the hands in your scoring pile separate by turning each set of cards at a right angle to the set below it. That makes it easy to count points and keeps the hands ready for the next round.

Endgame

When there are no hands left on the table, players can share with anyone who still has cards. On a shared hand, the first player to call Krypto scores the point.

After all the other cards are claimed, the player who has the last hand spreads it face up on the table. Everyone races for the final Krypto.

If all players agree it's impossible to make the target number with any hand, put those cards aside while you count score. Players score one point per hand in their scoring piles. Then return all ten sets of cards to the table, face down, to play the next round.

The player with the highest final score wins the game.

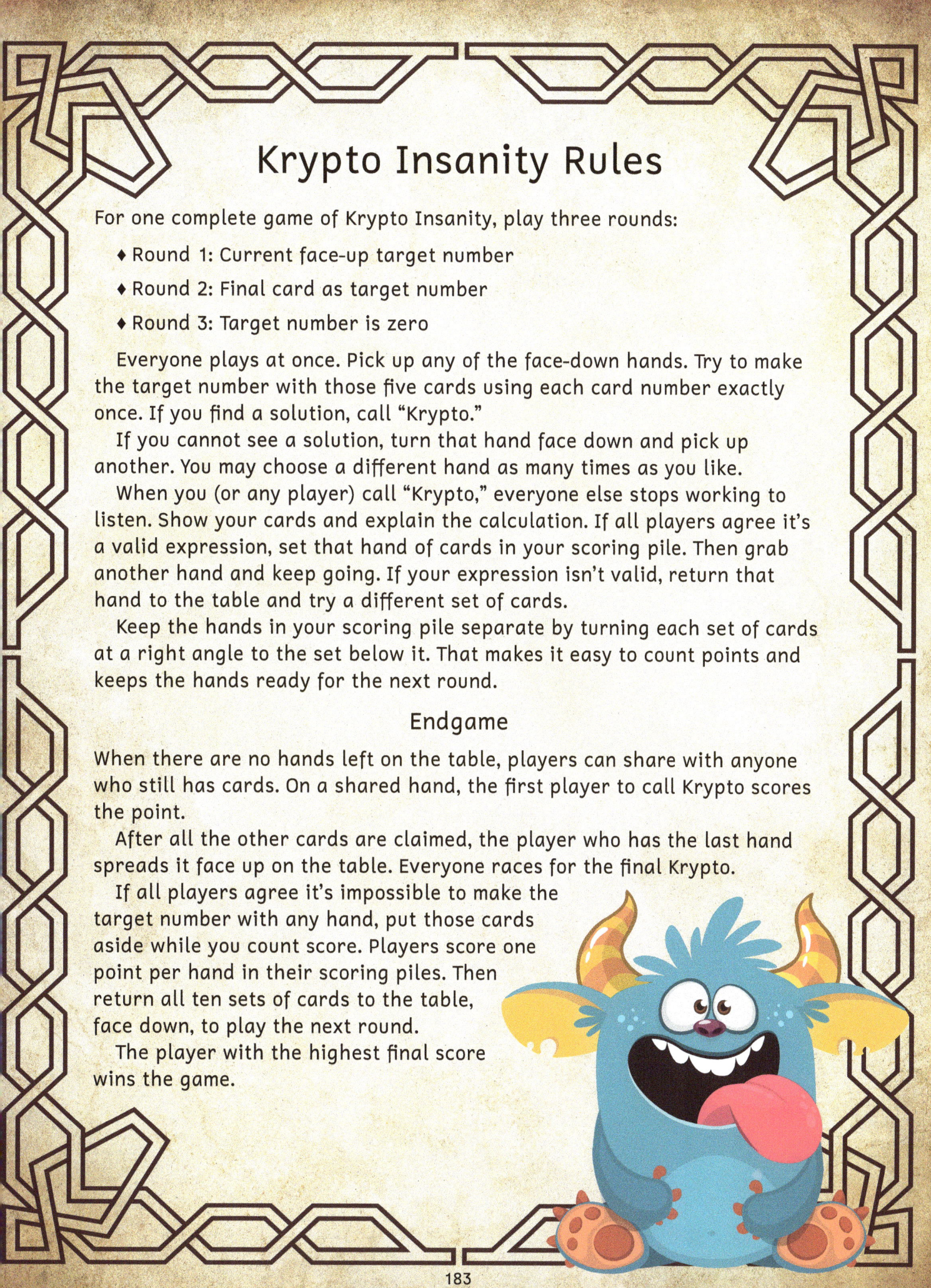

Notes & House Rules

Equation Master

Math Concepts: addition, subtraction, multiplication, division, order of operations, fractions, decimals, integers, absolute value, powers and roots, factorials.
Players: any number.
Equipment: one deck of cards, pencil and paper for each player and for keeping score. Calculator and timer optional.

Set-Up

Agree on which color represents negative numbers. Agree on the value of face cards—for example, jack = eleven, queen = twelve, king = thirteen. Or maybe you want all face cards to equal ten, while the ace can be worth the player's choice of either one or eleven.

Decide whether to set limits on the math operations allowed in your game. Or let dealers set their own limits at the beginning of each hand, before turning up any cards.

How to Play

The dealer turns up seven to ten cards (dealer's choice) on the table where all players can see. If desired, set a timer for ten or fifteen minutes.

Players write as many true equations—and if players have agreed to allow them, inequalities—as they can using the numbers on these cards.

Equations may involve any mathematical operation, and players are allowed to concatenate two cards to make a multi-digit number. Each card may be used only once in the calculation.

Endgame

When the timer rings or when players agree to stop, pencils go down and scoring begins.

Every valid equation or inequality scores 1 point per card number used. If no other player wrote the same expression, add a 5 point bonus for uniqueness.

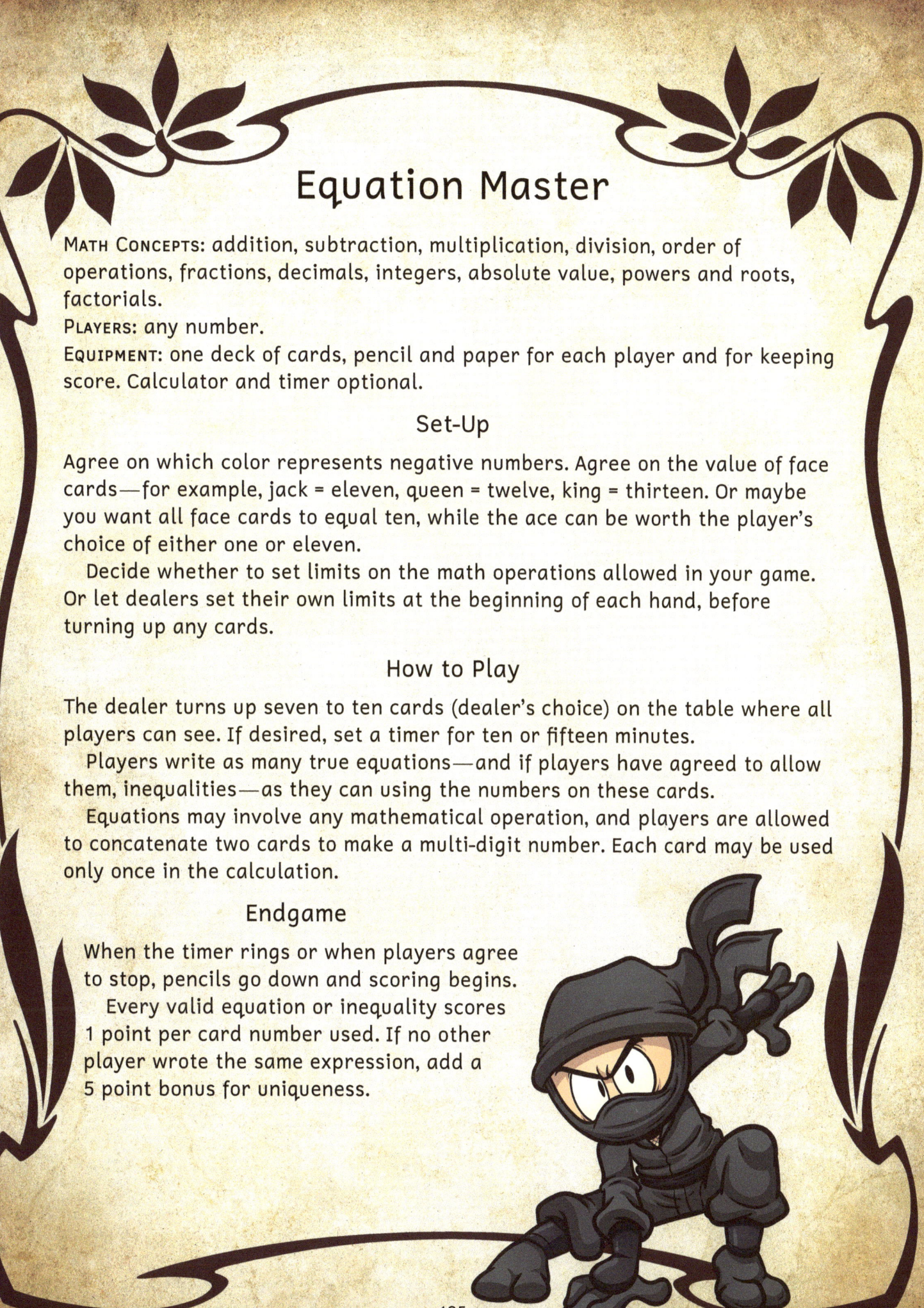

Notes & House Rules

Four in a Row Algebra

Math Concepts: integer multiplication, factors, multiples, algebraic multiplication.
Players: two players or two teams.
Equipment: gameboard, colored markers, and two paperclips, glass gemstones or other small tokens for the factors.

Set-Up

Choose a Four-in-a-Row gameboard for the players to share. Or have players work together to create their own gameboard:*

- On a blank sheet of paper, draw a 6 × 6 array of squares.
- In the space below the array, each player writes four factors. Factors may be positive or negative and may include fractions, decimals, and algebraic variables.
- Then take turns writing the product of any two factors into one gameboard square. A few duplicates won't ruin the game, but try to avoid writing a product that's already been used.

When all the squares are full, you're ready to play. You may want to slip your gameboard into a clear page protector or laminate it for playing with dry-erase markers.

How to Play

The first player places a paperclip on any factor at the bottom of the board. The second player places the other clip on a factor—the same or different—and then marks the product of those two numbers by coloring the square.

On each succeeding turn, a player shifts one paperclip to a new number and then marks the product of those two factors. If both players agree there are no possible moves, the player whose turn it is makes a fresh start by changing both clips.

Whichever player marks four (or more) squares in a straight row—horizontally, vertically, or diagonally—wins the game. The squares must touch each other at edges or corners, with no gaps.

If neither player connects four, then the player who has the most sets of three in a row wins.

*For free printable gameboards, download the Prealgebra & Geometry Printables at TabletopAcademy.net/free-printables.

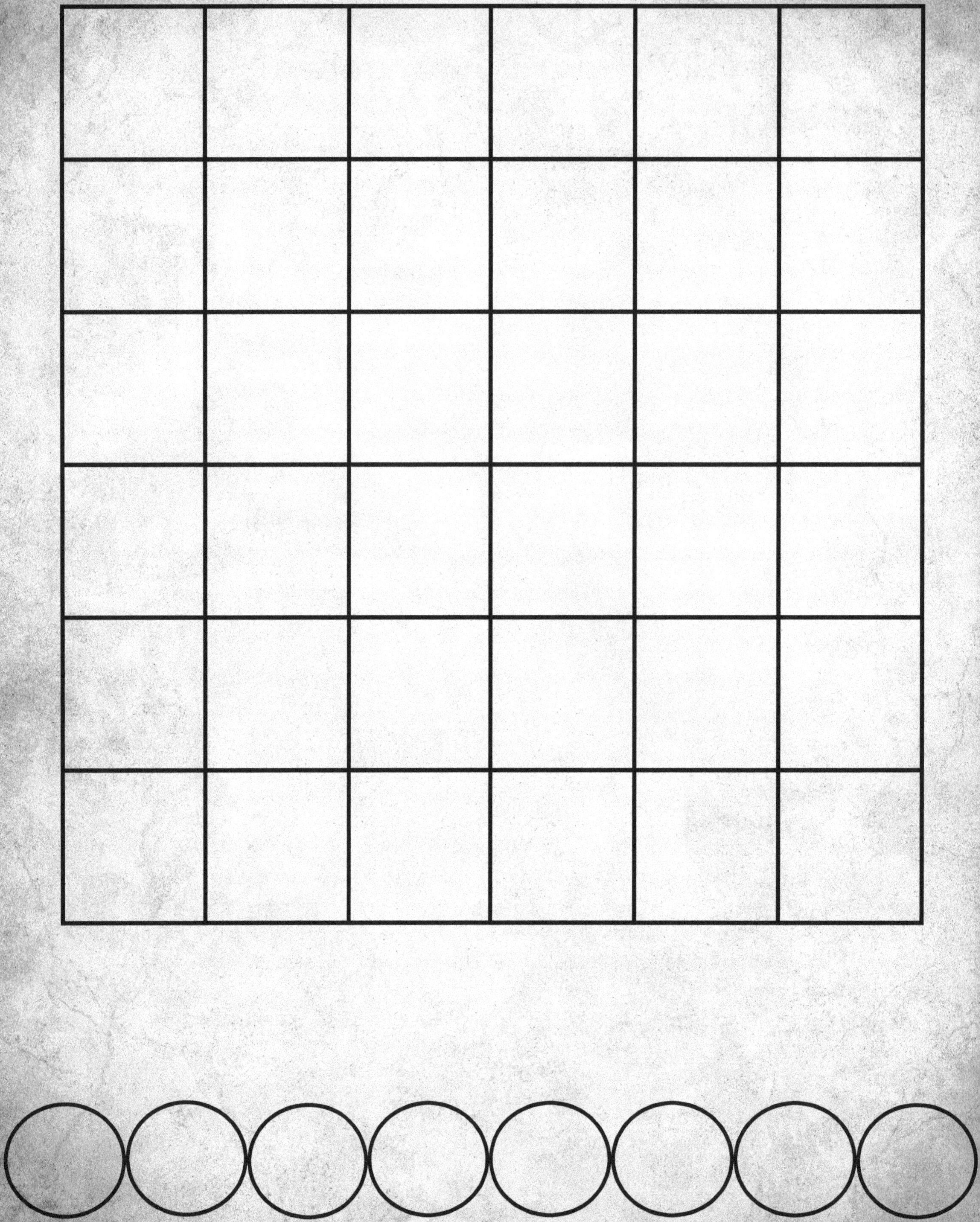

x^2	xy	x	$-2x$	$3x$	$-4x$
Ax	Bx	y^2	y	$-2y$	$3y$
$-4y$	Ay	By	1	-2	3
-4	A	B	4	-6	8
$-2A$	$-2B$	9	-12	$3A$	$3B$
16	$-4A$	$-4B$	A^2	AB	B^2

x^2	xy	x^3	xy^2	x^4	xy^3
x^5	xy^4	y^2	x^2y	y^3	x^3y
y^4	x^4y	y^5	x^4	x^2y^2	x^5
x^2y^3	x^6	x^2y^4	y^4	x^3y^2	y^5
x^4y^2	y^6	x^6	x^3y^3	x^7	x^3y^4
y^6	x^4y^3	y^7	x^8	x^4y^4	y^8

x^2	xy	$2x$	$-3x$	$4x$	$-5x$
x^2+x	$xy-x$	y^2	$2y$	$-3y$	$4y$
$-5y$	$xy+y$	y^2-y	4	-6	8
-10	$2x+2$	$2y-2$	9	-12	15
$-3x-3$	$-3y+3$	16	-20	$4x+4$	$4y-4$
25	$-5x-5$	$-5y+5$	x^2+2x+1	$xy-x+y-1$	y^2-2y+1

For More Information

Operations

Teacher Don Steward shared a series of "three operations" puzzles on his Median blog. I loved the puzzles and decided to turn his idea into a game.

♦ donsteward.blogspot.com/2020/01/three-operations-expressions.html

Exponent Pickle

John Golden invented the simpler "A × B^C" version of this game and tested it in his college algebra class. He writes:

"Older students also need these play experiences. I think they just abstract from them more quickly than younger students."

♦ mathhombre.blogspot.com/2017/07/same-game-different-grade.html

Krypto

Mathematicians have played with calculation puzzles (like the Four Fours) for centuries. Daniel Yovich invented Krypto in the 1960s and sold the game to Parker Brothers. The original version used a special deck of fifty-six cards and featured more restrictive calculation rules.

Four in a Row Algebra

I first saw this type of four-in-a-row game in 1987 on the old Square One TV show. Bill Lombard and Brad Fulton published the algebra gameboard in *Simply Great Math Activities: Algebra Readiness* from Teacher to Teacher Press.

◆ ◆ ◆

"I love games in general, but also in math class. We play many games as a family and love to bring in others to play.

"I think some of the reasons that mathematicians love math is that it is a lot like playing a game. Defined objects, rules declaring what moves are permitted, desired outcomes... a serious game that is."

—John Golden, *Games*

Algebra Match

A CHALLENGING GAME OF VARIABLES AND SUBSTITUTION

$x=y$ $x<y$ $x(y)=12$ $2y=8$

$x+y=7$ $3x>y$ $x=1$ $\dfrac{x}{3}=\dfrac{y}{2}$ x

$x+y>6$ $2y=x+2$ $x-y>0$ $y=2x$ y

$y-10=-x$ $y>4$ $x^2+y<6$ $x-2=y$

Notes & House Rules

Credits
Algebra Match is adapted from an algebra game by French homeschoolers Allison Carmichael and Martin Woods at the Parent Concept website: parentconcept.com/printable-algebra-game. John Golden invented the cooperative version of the game.

Algebra Match

MATH CONCEPTS: variables, substitution, algebraic equations, inequalities.
PLAYERS: two to four.
EQUIPMENT: blank index cards or card-sized pieces of paper, two six-sided or gaming dice, pencils or markers.

Set-Up

Share out sixteen index cards or blank squares of paper among the players. Players write an algebraic equation or inequality on each card, as follows:

- Use only two variables, x and y. Each card must use at least one of these variables.

- The values of x and y will vary based on a dice roll. Make sure your equation or inequality makes sense in this range of numbers.

- Create equations and inequalities that work with more than one value for x and y. For example, a card with "x = y" matches any roll of doubles.

Label one additional card "X" and another "Y." Save all these cards for future games, adding a few new cards each time you play, to create an Algebra Match deck.

How to Play

Lay all the algebra cards face up in a 4 × 4 array. If you have been collecting Algebra Match cards, shuffle your deck and turn up sixteen cards for the array. Pass the X and Y cards along with the dice for each player's turn.

On your turn, roll the dice. Find an equation or inequality card that fits your numbers. Choose one die to be your x value and place it on the X card. Place the other die on the Y card.

Explain how you know the card matches your dice, and then add that card to your personal score pile. You can only take one card per turn, even if more than one matches your dice.

If you cannot find a matching equation or inequality card, say "Pass." But if another player sees a match you missed, they may claim it. If you placed the dice on the X and Y cards before passing, the other player may switch them to make a match. With more than two players, the first one to speak takes the missed card.

The game ends when all sixteen cards are claimed or when three turns in a row end with a pass. The player who collected the most cards wins.

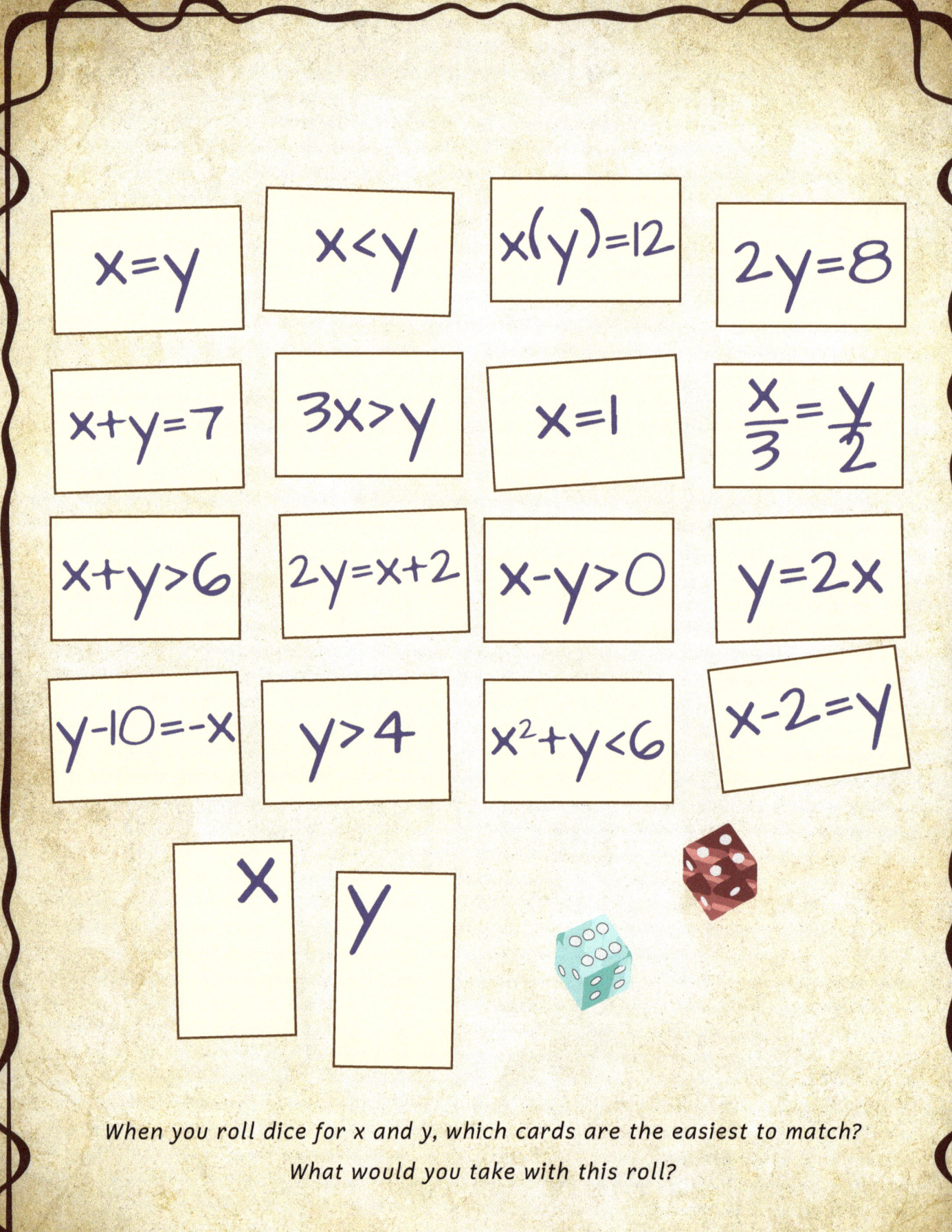

When you roll dice for x and y, which cards are the easiest to match? What would you take with this roll?

Algebra Match Variations

Cooperative Algebra Match

On your turn, if you can't find a matching card, you must choose any card to turn face down. If four cards get turned down, then the game is over, and all players lose. Players win by claiming all the face-up cards before a fourth card is turned down.

Algebra Multi-Match

When you have collected a large deck of Algebra Match cards, allow players to claim up to three matching cards at a time. Place new cards from the deck to fill the spaces in your 4 × 4 array after each turn.

Solitaire Multi-Match

On each roll of the dice, claim every card that fits your x and y values. How many turns does it take to collect them all?

Algebra-Tac-Toe

This requires just two players or two teams. You'll need a deck of at least thirty-six Algebra Match cards, plus two colors of poker chips or other tokens.

Arrange the cards face up in a 6 × 6 array. On your turn, place a chip on one card that matches your dice roll.

The first player to claim four cards in a row wins.

HOUSE RULE: Once a card is claimed, is it out of bounds? Or can the other player bump that chip and replace it with his or her own? Or perhaps you might allow both players to set chips on the same card? If players are allowed to bump others out or to share a card, then you may want to change the winning condition to five in a row.

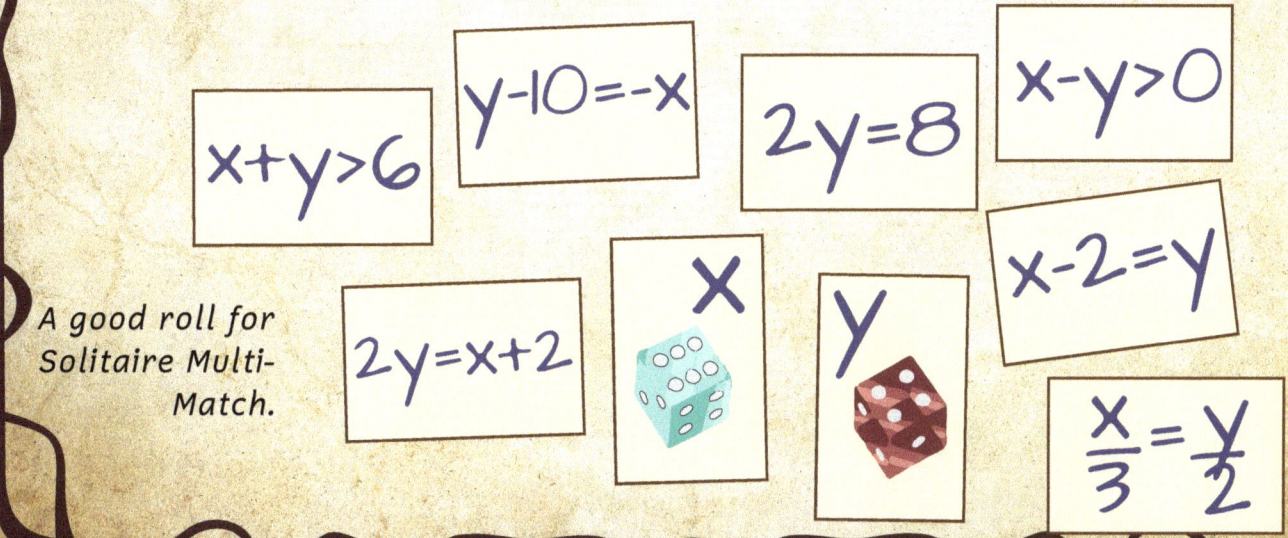

A good roll for Solitaire Multi-Match.

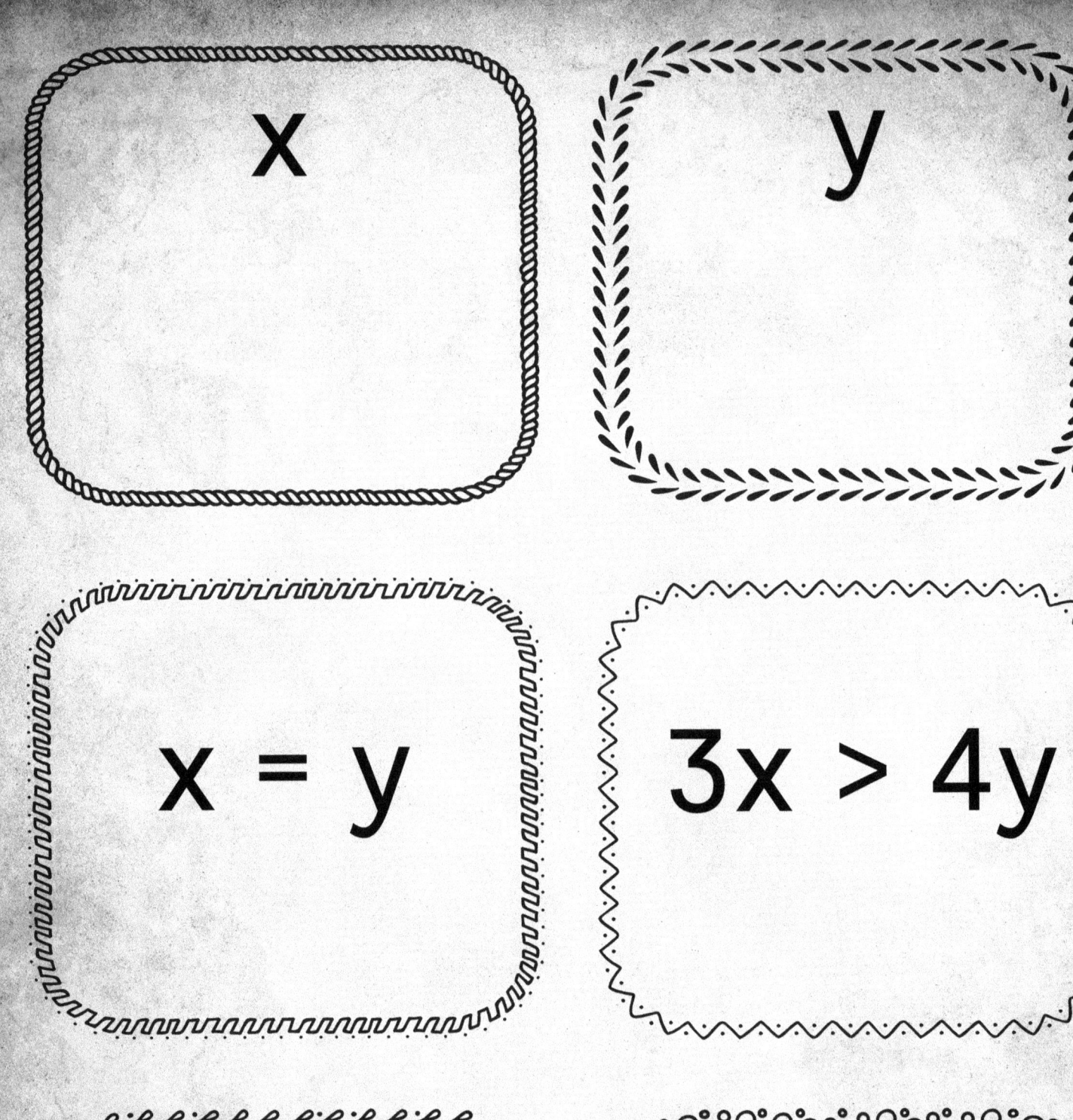

$xy = 12$

$2y = 8$

$x + y = 7$

$x - 3 > y$

$x = 1$

$\dfrac{x}{3} = \dfrac{y}{2}$

$x+y > 6$

$2y = x+2$

$x-y > 0$

$x = 2y$

$y-10 = -x$

$y > 4$

$x^2 + y < 6$

$x - 2 = y$

$x < 2y$

$5 = y$

$x < 7$

$x^2 > y^3$

Special Thanks

This book came to production with the help of many wonderful people who backed the project on Kickstarter. Your generous support and encouragement keeps me going!

I'm especially grateful to:

Alex	Julia Fay Deutsch	Peggy Jensen
Amanda Bartonek	Kayla Loeh	R. D. Peacock
Andrea Balderrama	Kirk's Tutoring	Rata Ingram
Blackard Family	Kristy Ramer	Rhys Kathryn Mathewson
Bronwen Dassing	LAD	Sasha Fradkin
Callister Family	Lisa Campbell	Sheryl Sotelo
Céline Malgen	Lizzy Petersen	Stan Halstead
Dan Finkel/Math for Love	Lucas & Izzy Dietrich	The Childers Family
David E.	Lynne Menechella	The Hancock Family
De Winter Tim	Margot Johnson	The Longbrooks
Dee Crescitelli	Melissa Bair	The Rineer Family
Heather C.	Meredith Clark	The Santos Family
Heather Daley	Michelle Scharfe	The Tully Family
Heather Hobbins	Murrieta Clark Family	Tracy Mickle
Ian Ashworth	Nacho Ruiz	Tracy Popey
Jeffrey Hellrung	Pattie Perry	Tulisha Jackman Scott
Jo Oehrlein	Paul P	Vic Beaumont
		Walter Thomas

Playful Math Books by Denise Gaskins

- If you're a parent trying to help your child learn math…

- Or a teacher looking for creative ideas for your classroom…

- Or a homeschooling parent hoping to enrich your student's understanding…

Then you'll love how Denise's books and activity guides lead you and your children to explore mathematics at a deeper level, building a strong foundation to support future learning.

Playful math enriches any curriculum. It doesn't matter whether your students are homeschooled or in a classroom, distance learning or in person.

Because everyone can enjoy the experience of playing around with math!

Tabletop Academy Press publishes playful math books and cool mathy merchandise for parents who want to help their children build the understanding and skills they need to succeed in school and beyond.

Homeschoolers, afterschoolers, unschoolers, and even classroom teachers appreciate our flexible approach that can work alongside any math program.

Visit us today:
TabletopAcademyPress.com

Or browse Denise's blog:
DeniseGaskins.com

www.ingramcontent.com/pod-product-compliance
Lightning Source LLC
Chambersburg PA
CBHW051331110526
44590CB00032B/4474